MINING IN WORLD HISTORY

GLOBALITIES

Series editor: Jeremy Black

GLOBALITIES is a series which reinterprets world history in a concise yet thoughtful way, looking at major issues over large time-spans and political spaces; such issues can be political, ecological, scientific, technological or intellectual. Rather than adopting a narrow chronological or geographical approach, books in the series are conceptual in focus yet present an array of historical data to justify their arguments. They often involve a multi-disciplinary approach, juxtaposing different subject-areas such as economics and religion or literature and politics.

Mining in World History

MARTIN LYNCH

REAKTION BOOKS

For my parents

Published by Reaktion Books Ltd
33 Great Sutton Street
London EC1V 0DX, UK

www.reaktionbooks.co.uk

First published 2002
Transferred to digital printing 2009

Printed and bound by Chicago Universtiy Press

British Library Cataloguing in Publication Data

Lynch, Martin
Mining in world history. – (Globalities)
1. Mining industries – History 2. Mines and mineral resources – History 3. Mining engineering – History
I. Title
622'.09

ISBN 978 1 86189 173 0

Contents

Preface

The idea for this book came to me during my involvement in writing *Back from the Brink: The Future of Australian Minerals Education*, which was published in 1998 by the Australian Minerals Council. It was apparent that the questions being asked by the study team were similar to those examined during the course of previous studies conducted over the last 100 years. It brought the point home that the history of an industry is frequently unknown to those working in it, yet the challenges faced and overcome by those who have gone before can provide inspiration and guidance to those working today. So I set out to write a concise and easily readable history of metal mining and smelting. It would focus on the main global themes as they have played out since the end of the Middle Ages. There already exist many excellent histories which can provide the interested reader with more details of a particular region or period.

It is one thing to set out to write a history, quite a different thing to write it, and over the last few years I have spent many hours in university and public libraries. In the course of this I have realized what a wealth of information is readily accessible to even the inexperienced researcher. This book is mostly the fruit of those researches. I have, however, received invaluable assistance in a few areas where information was not so readily available. I particularly want to acknowledge Marie-Christine Bailly-Maitre of the Centre National de la Recherche Scientifique, Grenoble, for her help in clarifying the history of the use of gunpowder in mining. I freely borrowed from the work of my father, Alban Lynch, in gaining a better appreciation of the history of mineral processing.

The most difficult part of writing is, I have found, writing clearly. For whatever clarity has finally emerged from the first drafts, I owe a good deal to the always constructive suggestions of Bob Burton of the Iron Ore Company of Canada as well as those of my wife, Pascale.

Prologue

This book is an attempt to recount and explain the story of metal mining and smelting as it has unfolded over the last 500 years. The full story, of course, goes back a lot further than that. The first smelters were probably simple clay bowls in which were heated very rich – almost metallic – samples of copper ore that had been chipped from rocky outcrops.[1] This practice seems to have had its origins somewhere in the triangle defined by the mountains of Armenia, central Turkey and the Sinai peninsula. As for the date of its commencement, some copper implements that are estimated to be 8,000 years old have been found in Turkey, but it is more likely that purposeful smelting first came into widespread and sustained use in this region some 6,000 years ago. It was once thought that the technology had spread from there to the rest of the world. More recent discoveries, however, have shown that in fact smelting was independently developed at later dates in centres as far apart as sub-Saharan Africa, south-east Asia, western Europe and northern China. It is possible, as Robert Raymond suggested in *Out of the Fiery Furnace*,[2] that the further development of smelting technology was linked to progress in the firing of pottery.

Copper was not the first metal known to the ancients. Meteoric iron was occasionally picked up from the ground. Also known was gold, which could readily be dug from sandy riverbeds, melted down and then fashioned into jewellery and ornaments. Copper was probably first used in the same way as gold. That is, it was employed mainly as ornament, for pure copper is too soft to be very useful in tools or weapons. But metal would not forever serve merely as ornament. It seems that our ancestors were continually tinkering with the smelting process and observing the effects of their experiments. By adding certain other likely looking minerals into the crucible they began to develop an understanding of alloying. In this way it was found that when a certain black sand, which could be found in some stream beds, was added to the copper ore in the proportion of about one in eight, the end result was a much harder and tougher metal that could be fashioned into implements that kept their sharp edge. The black sand was tin ore and the alloy was bronze.

With the advent of bronze, both tools and weapons could be made of metal, and from about 5,000 years ago the production of bronze became important enough to lend its name to a major epoch in the development of mankind – the Bronze Age. Bronze shields and cauldrons have been found in Europe, as also have moulds that appear to have served in the mass production of such objects as bronze dagger handles. The most prolific centres of manufacture seem to have been in northern China. The Shang bronzes, as they are known, are famous for their beauty and complexity. One specimen is an intricately decorated four-legged cauldron weighing nearly one tonne.[3] We may speculate on the size of the mining, smelting and metal-working complex that must have been in place to support the production of such objects. No-one knows from where was sourced the tin for the bronze smelters of the eastern Mediterranean. The best guess is that it had to be brought in from mines in Italy and Spain and, perhaps more importantly, south-west Britain.

At the same time as tin and copper were becoming commonly available, two other metals – silver and lead – also came into use. The renowned historian of metallurgy, R. F. Tylecote, would only go so far as to say that silver 'must have been known by about 2000 BC'.[4] This is the earliest date at which there are surviving silver objects. Nevertheless it was probably in circulation considerably earlier. A stone pillar from Mesopotamia dating from nearly 5,000 years ago is inscribed with a declaration on the prices of various goods. The prices are given in units of silver.[5] Whatever the date of its discovery, the appearance of silver represents another significant step forward in the metallurgical art. This is because it usually occurs in nature in close combination with lead (or with copper), and these metals had somehow to be separated during smelting. A process for achieving this was developed. It required the use of a saucer-shaped

hearth called a *cupel*, into which was placed the ore. This layer of ore was then subjected to smelting, and when it was complete the molten lead was poured off, leaving the silver behind in the bowl. So successful was cupellation that by 1500 BC silver was widely used as money throughout Mesopotamia and neighbouring lands.

From bronze the next step was to iron. Iron is harder and stronger than bronze and its ores occur in nature in much greater quantities than those of copper or tin. It has, therefore, considerable advantages over bronze. The Hittites, a people who came to prominence about 4,000 years ago in what is now central Turkey, were the first to possess iron weapons and tools.[6] The somewhat sudden appearance in their hands of these implements needs an explanation. After all, the metal itself had been known for at least a millennium before that. So why was it not more widely in use at an earlier date? The explanation is necessarily speculative, and rests on the fact that iron has one significant disadvantage when compared to bronze: it does not become molten until it reaches temperatures above 1,400 degrees Celsius, compared to the 900 degrees required for bronze. As the ancients could not achieve such temperatures in their furnaces, the iron they had to work with was a semi-smelted fused metallic lump that required much hammering and reheating before it was rendered as useful metal. Whenever there was sufficient bronze available, the production of iron may well have been avoided as being too labour-consuming.

It is probable that the Hittites used a furnace that did a better job of smelting iron than any before it. The design of furnaces had been in continual evolution since the first simple crucibles, one driving force behind the improvement efforts being the need to reduce the consumption of wood. Charcoal, derived from wood, was the only known furnace fuel and was scarce in the lowlands and flood plains that supported the first major civilizations. As the furnaces became more fuel-efficient they could also sustain higher temperatures. The adoption by the Hittites of iron-making may have been the result of their either borrowing or themselves developing a more advanced furnace design that made the production of iron a less laborious process. In addition to improvements in smelting technique, some

archaeologists have postulated a second factor in the rise of iron. They have speculated that the Hittites may have been driven to perfect their skills in iron manufacture during an era when invasions disrupted the trade routes along which tin travelled from the West.[7] In support of this speculation is the fact that the appearance of iron is roughly coincident with the period in which it is thought that Aryan peoples from the East invaded Europe, displacing its original inhabitants.[8]

The techniques for the large-scale production of iron seem to have travelled east along the Silk Road to China. Although they may have borrowed the technology from the West, the Chinese would take the art of iron-making to a height of sophistication that would only be surpassed in industrializing England. The first great breakthrough came sometime before 400 BC with the invention of double-acting bellows,[9] a device that provided a continuous blast of air into the furnaces. When water power was employed to drive these bellows, the furnace temperatures were able to reach the point at which iron was made 'to run like water'. And so the blast furnace was born. With the laborious task of hammering and reheating now at least partly done away with, production increased. The newly plentiful iron played a role in transforming Chinese society. Iron axes and iron ploughs accelerated the clearing of forest and made easier the proper ploughing of fields. Iron weapons gave Chinese soldiers an advantage over their enemies. The full benefits of this industrial transformation coincided with the coming to power of the Han dynasty. It was one of the most glorious periods of China's long history. Helped along by better-equipped armies and an expanded and more productive agricultural base, the Chinese empire grew in unity and power.

At the same time as the Han reigned in China, the Romans held sway over the lands bordering the Mediterranean. Their most productive iron regions lay in the western provinces, and iron-making towns have been identified along a line stretching from the middle of what is now France up to Germany.[10] The works were not as large or technically advanced as those in China, and indeed the formidable technological breakthroughs that occurred in China seem never to have travelled west. The production of molten iron was not mastered. Even so, water-

powered machinery was in common use for driving the bellows, and the total quantities of metal produced were substantial. Iron was of course not the only product of the Roman smelters. Silver was needed for both coinage and plate, and lead was produced in vast amounts for use in lining aqueducts and piping water into the houses of the well-to-do. It was also used as a coating for cooking utensils, in the production of wine and in the preservation of food[11] – all of which uses have been seized on from time to time to support the discredited claim that the Roman empire declined because of a chronic lead poisoning of its citizenry. A major source of silver continued to be the fabulous mines of Laurion in Greece, still productive despite the fact that they had been intensively worked for many hundreds of years. Spain was the region most richly endowed with ores. It was home to the copper and silver mines of Rio Tinto, probably the most productive and well-equipped of Imperial Rome. It was also the empire's most prolific producer of tin and lead, although here it faced tough competition from the misty island of Britain. In the south-west corner of that island, tin and lead were found in abundance.

Let us now jump ahead to AD 1000. After a period of damaging civil wars, Imperial China was once again prospering. The emperors of a new dynasty, the Sung, had restored order and brought about much-needed administrative and tax reforms. The Great Wall had been completed and now formed a lengthy defensive road separating China's northern provinces from the marauding horsemen of Mongolia and central Asia. The production of metals was flourishing. Copper coinage had come into common use and silver production was being encouraged, in part because the emperors had adopted a policy of using it as tribute to slow the encroachment of the northern barbarians. Saltpetre was mined to provide the raw material for gunpowder. While this black powder was mostly devoted to military purposes, it is probable that it was also used in mines as an explosive to break the rock.

The first Sung emperor had chosen for his northern capital and principal garrison K'ai Feng, a city that lay on a tributary of the Yellow River, the largest of the many large rivers that run through northern China. Following the emperor's declaration,

the town's population swelled dramatically. Court officials, bureaucrats and soldiers all took up residence. It is estimated that at its peak the inhabitants may have numbered one million. Such a city had an enormous appetite for metal, and soon blast furnaces were dotting the surrounding districts. It is said that the construction of the royal palace drew in such quantities of iron and wood that it placed a severe strain on the resources of the regions nearby. Wood was already scarce, and in time it became expensive. Increasingly the populace turned to coal for heating their homes. No doubt the incentive became strong to use coal for firing the blast furnaces too, because iron-making is a glutton for wood.

Coal had been used in certain regions as the furnace fuel as far back as the Han dynasty. Its successful use in place of charcoal was no mean feat, as we shall see in a later chapter. It is a measure of the continuing sophistication of Chinese metallurgy that towards the latter part of the eleventh century, coal-based blast-furnace iron-making was mastered in the mountains around K'ai Feng.[12] This plentiful and inexpensive fuel seems to have unleashed a surge of expansion, and by the turn of the century K'ai Feng was being borne along on a tide of iron. The smelting works grew large. One of them is recorded as having produced 14,000 tonnes in a single year. This was more than was being produced in all of contemporary Europe. No doubt much of it was devoted to the manufacture of swords, armour and other weaponry. If so, it would soon have been put to good use, for mounted invaders were once again battering against the Great Wall. Their successors, the Mongols, would not stop before Genghis Khan was emperor of all China. The massacres and famine that accompanied these invasions meant the end for K'ai Feng. Caught up in the maelstrom was the industrial complex that supported the great city. The blast furnaces were snuffed out.

Even in more peaceful times the production of metals in China had come under threat. An Imperial Order banning mining activity was issued in AD 1078. While it was widely ignored, the Order reflected the views held by many Chinese. The Confucian ethic frowned on an activity that removed the limited resources of the earth in such a fashion. Worse still, the

mines were a magnet for the kind of corruption that could end up causing riots and civil disorder. Corruption, indeed, was the curse of Chinese miners. A Commission that had been instructed to report on the affairs of the mining industry was moved to write:

> Nature has provided us with excellent deposits. These deposits were capable of producing much profit to the people. The officials, thinking that there was very much money in mining business, wished to take it for themselves, so that in every mining district corrupt practices grew up amongst them, to the very great injury of the people. For this reason the rich refuse to devote their capital to mining and mining enterprises are gradually ruined. If a capitalist puts his money into mining, before he has gained a profit evil characters raise complaints against him, with the result that the officials banish him to some distant country or take possession of his belongings, in spite of his complete innocence. It will thus be seen that Chinese mining affairs are exceedingly badly managed.[13]

Many in the ruling class thought that the mines were not worth the trouble. The Mongol khans who replaced the Sung emperors shared this view. They allowed them to decay, along with the other industrial advances of their conquered subjects. That great stimulus for mining, the need for currency, disappeared altogether. The Mongols instead used a paper-based money.[14] Under their stewardship, the momentum of Chinese metallurgy would be lost. Later dynasties would make no great effort to recover it.

At the other end of the Eurasian land mass, a different course of events was in train. Some 500 years before, the western Roman empire had collapsed amid invasions by Germanic peoples from the north. Across Europe an extended period of economic hardship and stagnation ensued, characterized by subsistence farming, lawlessness and continual warfare. A succession of strong rulers gradually restored order, until finally the peoples of western Europe seem to have mastered their situation sufficiently to begin a general movement eastwards in a search of more, and more fertile, farming land. The German

peoples were at the forefront of this migration, and many pushed into the Slavic lands (Bohemia, Hungary, Poland). As they fanned out into the new lands, some brought with them prospecting skills; the rest came equipped with fresh eyes and their own natural resourcefulness. The first major mineral discovery came in the rugged and heavily forested Harz mountains of northern Germany. Tradition has it that the great silver and lead deposits of the Harz were found by a nobleman who was out hunting. Having tethered his horse, Ramelius, to a tree, the nobleman set off on foot. On his return he found that the horse had pawed the ground and exposed the shining black ores that were soon to yield a fortune in silver and lead. Within a few decades silver was pouring out of Rammelsberg and the nearby town of Goslar in quantities unheard of in Europe since Roman times.[15] Rich ores were next found in the Alps of eastern France and in the mountains of the Tyrol.

Then, sometime around AD 1160, salt merchants from the eastern German town of Halle, who had been selling their cartloads of salt blocks in Bohemia, found their way hindered by a wash-out of the cart track along which they were travelling.[16] The track at this point was running through the foothills to the north of the mountains that separate eastern Germany from Bohemia. Noticing a shiny black ore similar to those they had seen when plying their wares in the Harz mountains, the merchants took some samples and had them tested at Goslar. The ore was richer and yielded a finer grade of silver than the Harz. It was a bonanza. As the news of the richness and extent of the discovery spread, experienced miners and others willing to try their luck flocked to the site. A *Sachstadt* ('town of the Saxons') soon sprawled on the mountainside. Some flavour of this settlement may be taken from the name given it. It was called Freiberg, the Free Mountain. In its wake came a flood of other finds – of silver and gold as well as of copper, lead and tin – the length and breadth of the continent. The silver and gold was greedily sucked into the coining mints, as in the wake of agricultural improvements commerce was becoming more vigorous and coinage was in demand.

Large strides were also being made in the technology of iron production, although there seems not to have been any influence

from the iron-works that were at that time thriving in China. The medieval iron-maker's struggle to achieve sufficient furnace temperatures had led to the adoption of the *Stuckofen*, a furnace design that due to its height and greater capacity was able more readily to sustain high temperatures, to produce larger volumes and to use fuel more efficiently in doing so.[17] Even so, the best that could be done was to produce a fused lump of iron, charcoal and other impurities. The job of the forge worker, as it had been in the time of the Hittites, was then to reheat this lump while continually striking it with a hammer in order to drive out the charcoal residue and other impurities. The result was wrought iron. When the long-neglected water-wheels of Roman times were rediscovered, the *Stuckofen* grew in size. Not only could water drive the larger hammers that were needed to beat the *Stucke*, it could also drive the bellows day and night.

At the forefront of these developments were the German master miners. Having gained experience in the Harz and Saxony, these men travelled to the new mineral fields across Europe, transferring knowledge and skills. They were needed, for example, to organize the digging of drainage tunnels, where flooding was becoming a problem in the underground workings. They were also needed to construct and operate water-driven stamp mills to crush the larger pieces of rock and water-driven bellows to fan the furnaces. Their skills in operating the smelters were much sought after. Such was their reputation that Stefan Uros, King of Serbia, gave them the task of opening up his Brskvo silver mines.[18] At about the same time the rich silver outcrop of Iglesias was discovered in Sardinia. Sardinia was controlled by the powerful Italian trading city of Pisa, and there was never any doubt that these wealthy merchants would delegate the exploitation of their bonanza to the best people they could find for the job. Thus, Germans travelled to Sardinia. This prosperous scene would last until the middle of the fourteenth century before decline set in, brought about by a combination of war, the ravages of the Black Death, wider economic decline and the depletion of the known mineral fields. To begin the task of tracing the path that led to the mining and smelting industry of our day, let us now turn to examine the revival that followed this decline.

ONE

The Metallurgical Renaissance

LIQUATION

The German master miners had long known that the copper ores of central Europe contained abundant silver. What they did not know was how to get their hands on it. It comes as no surprise, then, to learn that the invention of liquation – a process for liberating silver from copper – had the effect of reviving the ailing mineral fields of the region. Liquation appeared at the beginning of the Renaissance, a period of European intellectual awakening that originated in the palaces and grand households of the wealthy Italian city-states. The Italian princes and merchants were the middlemen between Europe and the Orient, and had prospered on the trade that flowed through their ports. Now their money, their interest in the forgotten works of the Greeks and Romans and their search for artistic magnificence provided the support for the new breed of scholars and artists to ply their trade. But when liquation first emerged into the light of day, the new thinking of the Renaissance would scarcely have had time to move out of Venice, Florence and their sister cities. Thus, liquation was not the product of the intellectual ferment of the new era. So whence came this technological leap forward?

The first documented case of the use of the process has been found in the archives of the municipal foundry of Nuremberg. The year was 1453.[1] Its appearance in Nuremberg is not surprising. That city had for many years been one of Germany's main centres of metal refining and fabrication. Copper and brass goods and all manner of metal utensils were made there using the ores that arrived in ox-drawn wagons from Saxony and Bohemia to the north-east. It was home to one of the largest mints in Germany. Already Nuremberg had a reputation as a leader in metallurgical techniques.

Understanding the route of discovery of the *Saigerprozess* (the German name for liquation), as opposed to the location of its first use, requires a bit more guesswork. The metallurgical science of the day was crude indeed, as it was in all other areas of technology. One commonly held belief was that metals grew in the ground. The scientific method of observation and experiment did not yet exist except perhaps in matters of astronomy, and laboratory experimentation, if we may use the term loosely, tended to be the preserve of the alchemists. 'The alchemists'[2] is a convenient term for the widely diverse group of people who worked with metallic ores and all sorts of other natural substances, seeking to distil the essence of these materials both singly and in combination. They claimed to be able to change substances from one state to another. Lead, to take the clichéd case, could be 'transmuted' into gold if only the right ingredients were added and the appropriate words uttered. It was a shadowy pursuit, and the alchemists tended to obscure their practical endeavours within a thicket of mystical writings.

By the time of the Renaissance many had come to view alchemy as a dubious practice, and its precepts were the subject of debate amongst learned men. A grudging supporter was Vannoccio Biringuccio, an Italian metalsmith of some influence in the city-state of Siena.[3] It seems that the Biringuccio family enjoyed the patronage of the dictator of that city, and Vannoccio Biringuccio became a master at the casting of bells and armour, eventually putting his lifetime of observations on smelting and related matters down in a book, *Pirotechnia*. His was a tempestuous life – twice he was exiled from Siena following popular revolts – and it is perhaps this background that explains the combative tone in which from the pages of *Pirotechnia* he denounces the alchemists. 'How many alchemists have I heard lamenting, one because by some unfortunate chance he had spilled his whole composition in the ashes; another because he had been deceived by the excessive strength of the fire … and yet another because he had poor and feeble materials!'[4] He makes a telling point by asking why, when they have such powers freely at their disposal, the alchemists were usually poor men. Yet in a different part of the book, his diatribe having run

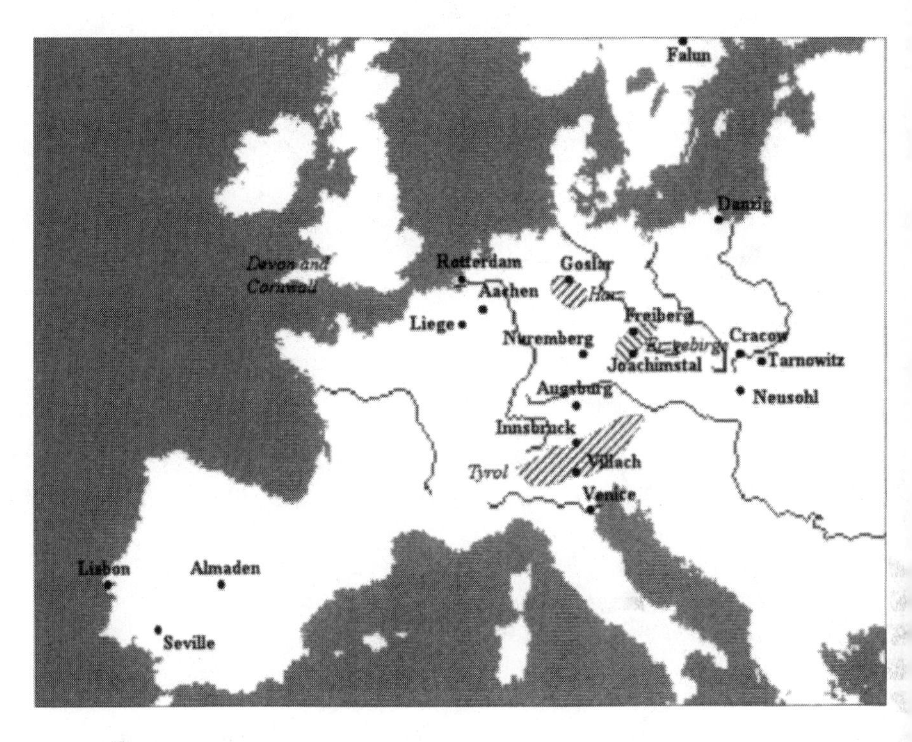

1 Europe, 1550

its course, Biringuccio concedes that the practice of alchemy is, despite everything, a productive occupation that 'gives birth every day to new and splendid effects such as the extraction of medicinal substances, colors, and perfumes, and an infinite number of compositions of things'.[5] It is likely that liquation first saw the light of day in the sombre surrounds of an alchemist's workshop.

The next step was for a coppersmith to try his hand at putting the new process to practical use. Where better for this to be done than in Nuremberg, where there existed the skilled workers, many foundries and plentiful ores of all kinds? Soon there existed in Nuremberg five liquation works, and within another fifteen years the process had spread throughout Germany, Poland and the Italian Alps. As it spread, so European metal production leapt upwards. How did the process work? Essentially, it relied on the fact that when a furnace packed with a mixture of metallic lead- and silver-bearing copper ore was smelted, in the process of cooling down, the silver contained within the ore would be absorbed by the lead. The result was a congealing mass of molten metal, in which the silver had moved from the copper to the lead. The next step was to separate the lead from the copper. This was done by taking advantage of the fact that lead has a melting temperature far below that of copper. Large plate-like discs – called liquation cakes[6] – were fashioned from the copper–lead mix and allowed to solidify. They were then placed in a special furnace and heated until the lead melted and drained out of the cake, leaving the still solid copper behind. The molten droplets were drained away in a gutter placed underneath the furnace grate. Once this was done it was straightforward to remove the silver from the lead using the time-honoured process of cupellation.

Performed crudely, liquation was very inefficient and hardly worth the cost of the lead that had to be purchased to allow it to proceed. This is one reason why the process tended to be operated on a large scale. The typical building in which it was housed measured, according to one contemporary, 50 x 20 yards. Small-scale 'backyard' operations would not allow the careful control that had to be exerted to make the process a success. So tricky was it to operate that the technical guidance of the *Saigerprozess* tended to remain for long in the hands of the experienced master metalworkers from Nuremberg. Under their skilful guidance the moribund mining districts of central Europe came once more to life.

The first regions to prosper as liquation began to spread were Schneeberg in Saxony and Schwaz in the Tyrol. Then the mines around the town of Annaberg came into production, and not long after came Marienberg. Both were situated in the Erzgebirge, the mountains separating Saxony from Bohemia. In 1516 was found the biggest of the Renaissance silver strikes, Joachimstal. Its fame and prodigious production became such that it would eventually, via the coin known as the Joachimstaler, give us the word dollar. In addition to the new finds, liquation helped to revive the older fields. In this way Rammelsberg and Freiberg became once again a hive of activity, as did some of the abandoned fields in Bohemia. All of these enterprises needed capital to sink the wide shafts and to build machinery and smelters, and so the new driving force in the industry was the rich merchant. Mining attracted some of the greatest merchants of the age, including the French royal banker Jacques Coeur and the Medici, the Florentine family of bankers and popes. Mostly, however, money for mine and smelter construction flowed in from the towns.[7]

The merchants of Augsburg, situated next door to the Tyrol and the Austrian Alps, funded much of the activity in those regions. The rich burghers of Nuremberg and other large German towns did the same in the mountains of Bohemia and the Harz. Like investors of any age they were after quick returns on their money. And quick returns there were to be had. At the time Europe was experiencing something akin to a copper shortage, and prices were buoyant. There had been a rise in the demand for all the traditional uses for copper, and many new uses appeared. Ships were becoming larger, more numerous and more sophisticated. Their instrumentation was fabricated from copper, as were their coverings and screws. Building fashions, especially roofing, now required copper, where once lead had been used. European armies came to depend more and more on artillery when besieging opponents' strongholds, a development that meant a dramatic increase in the production of brass cannon. It comes as no surprise, then, to learn that when word got out of a rich copper strike, investors clamoured for a piece of the action.

The liquation works were hungry for lead and initially were held back for the lack of it. So when a large lead-bearing outcrop was found at Tarnowitz, near the Polish city of Cracow, rapid development was soon underway. One of the men to drive it was Jan Thurzo. In addition to financing the sinking of mine shafts he built a large *Saigerhütten* in Cracow, a city situated on the road between Hungary and the Baltic port of Danzig.[8] Thurzo's smelting works became one of the principal industries of the city, taking the Hungarian ores and transforming them into copper and silver that were then sent on to Danzig for export to Holland, France and England. Such was its consumption of wood for fuel that the works was likened to Mt Etna because of the clouds of smoke that billowed from its furnaces. A special road was built across the Carpathian mountains so that the transport of the metals could proceed more rapidly. It was known as Thurzo's Road. Other *Saigerhütten* were built in the town of Villach at the foot of the Bleiberg. In a similar way to Cracow's position on the trade route north, Villach was nicely situated on the route south from Germany and Hungary to the port of Venice. The lead of Bleiberg and Tarnowitz was also exported to Nuremberg, Aachen and Cologne, where it was used in the *Saigerhütten* of those cities. In this way, a network of metal-trade routes began to criss-cross Europe.

If the liquation industry was the highest achievement of the metallurgical art of the time, the iron works were not far behind. They had become formidable industrial complexes. The bellows had grown so powerful that now molten iron could be produced at will. There is a record of a blast furnace in the south Netherlandish town of Namur in 1410.[9] Four decades later another is known to have been in operation in nearby Liège. As the Chinese had known well, the product of the blast furnace operated in this way will be brittle. A sharp shock can cause the iron to shatter. Its uses were, therefore, somewhat limited. Nevertheless, by the time of the Renaissance some important applications for this 'cast iron' had been found. Molten bronze had long been poured directly into moulds in order to produce church bells, and the technique had been transferred to the fabrication of bronze cannon.[10] In the late medieval period, smelter men had discovered that iron, too, could be poured directly into moulds to make bells and, much more usefully, cannon.

The cast-iron cannon ordered by England's Henry VIII for his Continental battles were manufactured near Liège, a town that lay at the heart of a thriving heavy industrial region. The easily accessible coal deposits in the surrounding hills were exploited to provide fuel that, although it could not be used in the blast furnaces, could be effectively employed in the subsequent refining steps required for the fashioning of fine weapons and armour. It was a first for Europe. From Namur, just down-river, skilled furnace operators migrated to northern France and across the Channel to England, taking with them the secrets of cast-iron manufacture. If King Henry had wanted brass cannon instead of the less expensive iron variety, he could have found them in Liège too. For that town, along with Aachen to the north, had become an important centre of brass-making thanks to the discovery nearby of large and easily worked calamine deposits (brass is an alloy of copper and zinc; calamine is a zinc-bearing ore).

THE NEW WORLD

Before we look more closely at this evolving industrial structure let us turn our attention to a different scene. It was during the Renaissance that for the first time in several thousand years the sailors of the Mediterranean began to venture out into the rougher waters of the Atlantic. In part this was the natural outcome of their awakening curiosity, and in part it was because their once-lucrative trade in spices and silks from the East was now blocked by the powerful and hostile Ottoman Turks. The sailors who ventured into the Atlantic were searching for an alternative route to the Indies. In addition to the abundant spices and silks, there was gold to be had. Had not Marco Polo told of an island empire off the coast of China possessing 'gold in the greatest abundance, its sources being inexhaustible'?[11] Marco Polo had gone on to say that the enormous treasure had inspired Kublai Khan to attempt two invasions. Here was a prize worth striving for, especially given the persistent shortage of gold in Europe.[12]

The Portuguese led the way, mounting a series of exploratory expeditions down the coast of Africa. Logic told them that if a ship followed the west African coast for long enough, it must at some point reveal a route to the East. After half a century of heroic adventure they had got as far south as the Guinea coast and there established a thriving trade with the African tribesmen who, in return for textiles, mirrors and slaves, exchanged the gold they brought with them from the interior. A fort was built, São Jorge da Mina (St George of the Mines),[13] the first permanent settlement to be established by Europeans in sub-Saharan Africa. The sea captains continued to push south, and in 1488 Bartolomeu Dias rounded the Cape of Good Hope, camping on the banks of the Fish River before returning the way he came. The next expedition, ten years later, was led by a Portuguese courtier named Vasco da Gama. After a twenty-month round-trip, da Gama sailed into Lisbon with a fleet laden with spices and other precious goods from the Indies. His arrival caused a sensation.

The Portuguese successes were surely at least part of the reason why the young monarchs of Spain, Isabella and Ferdinand, decided to support the Genoese sea captain Christopher Columbus in his quest to find yet another route to India. Columbus proposed the radical idea of heading west in order to reach the East, taking advantage of the fact that the world is a globe. Equipped with three small ships, in August 1492 he duly set out from the Spanish port of Palos and sailed away over the horizon. Seventy-one days later he landed on the Caribbean island he named San Salvador. After some island-hopping he eventually established a settlement on Hispaniola (Haiti). In one of the first journal entries of his epic voyage Columbus described the purpose of his expedition as being principally to oblige the Great Khan of the Indies in his request for instructors in the Christian faith. Probably more heartfelt, however, were the numerous and repeated observations about gold and spices. Soon after his arrival he recorded that in Hispaniola 'there are many spices and great mines of gold and of other metals.' Later on: 'In another island, which they assure me is larger than Hispaniola, the people have no hair. In it, there is gold incalculable…'.[14]

Columbus was to be disappointed. By the time he set sail for home, of spices he had found little, while Hispaniola's creeks and gorges had yielded but a small amount of gold. In all, he was to make four voyages to the West Indies, and on each he lifted a bit further the veil of mystery that hung over this new land. He never did visit the court of the Great Khan. Instead, he installed his unpopular brothers as rulers over the fledgling Spanish settlements. Their leadership, and that of Columbus himself, was so autocratic and divisive that complaints were made to Ferdinand and Isabella. They appointed Francisco Bobadilla to investigate, and in 1499 the discoverer of the New World was escorted home in leg-irons. The work of establishing Spanish dominion in the Americas was to be taken up by others.

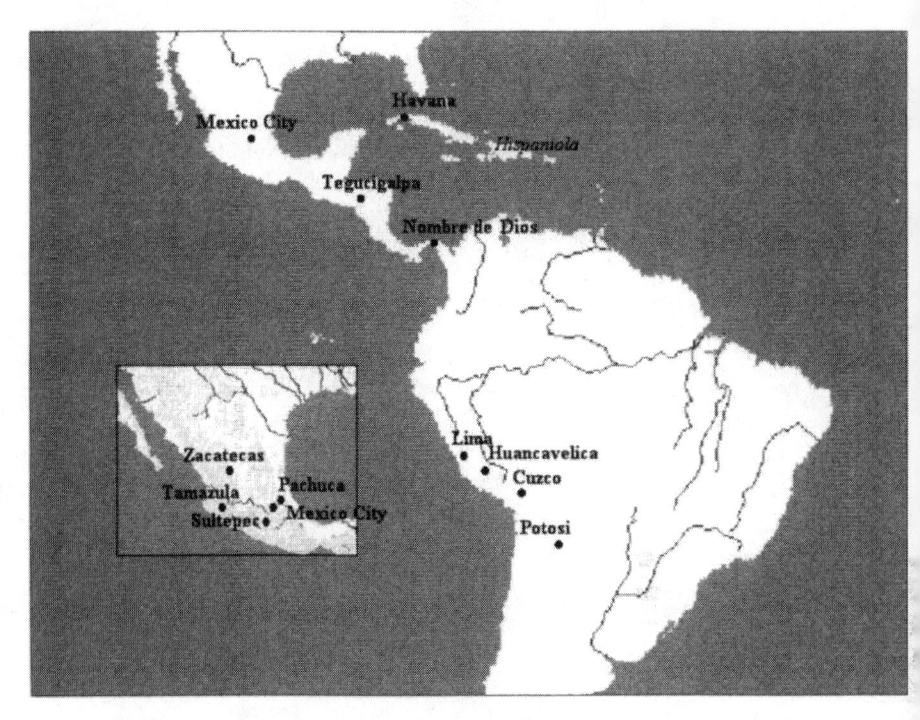

2 Spanish America, 1550

The real beginnings of Spanish settlement in the Americas date from the appointment of Bobadilla's successor, Frey Nicolas de Ovando. Under his harsh regime the production of gold from Hispaniola increased steadily. It was dug from the rivers and streams of the island using the forced labour of Indian workers. The Spanish had no mining tradition, and they were not disposed to employ expert miners from central Europe. Instead they simply drove the Indians beyond the limits of their endurance. It was a wasteful and brutal business.[15] Rules were established to manage the growing industry. The city of Seville was to be the only port through which trade with the colonies could be conducted, and the Casa de Contratación (House of Trade) was established there to receive the New World bullion. Any precious metals found were subject to taxation at a rate of one part in five. Gold smuggling thrived. As production increased so did the Indian death rate, and it was not too long before Hispaniola was experiencing a labour shortage. Captives from West Africa were shipped in to provide slave labour to supplement the dying Indians. The supply was never enough and, prompted by the labour shortage as much as by any desire for further discovery, the Spanish began setting out for neighbouring islands. First Jamaica was colonized, then Cuba, then Puerto Rico. Cuba came to rival Hispaniola as a source of gold.

In time the Spanish gaze turned increasingly towards the mainland, until finally Hernán Cortez mustered a force of perhaps 600 men and set sail west from Cuba.[16] He had the vague goal of establishing trade links with the powerful tribe that was rumoured to control the mainland. Cortez was a ruthless opportunist and a daring leader, but still there was nothing in this prosaic beginning to suggest that within the space of some 30 years his band, and a second led by Francisco Pizarro, would have conquered two mighty and sophisticated civilizations, and that Spanish soldier-adventurers would hold sway over the Americas from the Rio Grande to the Andes. Cortez and his men overthrew the Aztec empire of Mexico, utterly destroying the inland capital, Tenochtitlan. Pizarro and his company did the same to the even stronger Inca civilization.

This sophisticated people controlled an area stretching along the Pacific coast for more than 3,000 miles from modern Colombia down to central Chile, and reaching inland from the coast to the forests of the Amazon. Pizarro's conquest was spurred on by an obsessive greed. He had convinced himself that the civilization he had encountered was awash with gold. The captured Inca emperor recognized this obsession, and offered as his ransom to fill full of gold and silver a room of about 100 cubic yards in volume. It netted the Spanish six tonnes of gold and twice as much silver, but far from placating them it only served to feed the fire of their rapacity. Accusing the imprisoned emperor of treachery, they burnt him at the stake and went on to crush resistance through the rest of the empire. In 1533 the Inca capital of Cuzco fell, reportedly yielding even more precious metal than had the ransom.

In the wake of the plunderers came the administrators. Lands were allotted, rules established for the treatment of Indians and a system of government put in place. The mainland was divided into three provinces, or Vice-Royalties. The northernmost was New Spain, which occupied present-day Mexico and the central American states. Venezuela, Guyana and Colombia were lumped together to form New Granada. And sprawling along the Andean mountain chain was the Vice-Royalty of Peru. Up and down this vast expanse the hunt for treasure went on. Nothing was so sacred that it could not be pillaged for gold and silver. Temples, palaces and even graves were ransacked. In fact, however, neither the Aztec nor the Inca societies were great storehouses of gold and silver objects. In Mexico, in particular, the booty that fell to the conquerors was disappointingly meagre. Thus thwarted, the Spanish turned to chasing rumours. El Dorado reputedly lay somewhere to the south of Panama, while the wondrous Seven Cities of Cibola, it was said, were somewhere to the north of Mexico City.

The incredible speed with which great tracts of Central and South America were explored is testament to the fervour with which the Spanish sought the elusive treasure. Some discoveries were made. Gold was found in the sandy streams in several places throughout Mexico. Further south, rich finds were made at Tehuantepec and then Tegucigalpa, now capital of the Honduras. The use of Indian slave labour to dig the gold – legal under Spanish law because they had been captured in war – was widespread. As they had on the islands, these unfortunates died in droves, victims of European diseases and of the unremitting heavy labour for which they were totally unprepared. While none of the finds was truly spectacular, by the 1540s gold was flowing in gratifying quantities back to Seville and the treasure from the Americas had become an important source of revenue for the Spanish crown.[17] That crown now rested upon the head of Charles V of the Austrian House of Habsburg.

MAXIMILIAN AND THE CAPITALISTS

The kingdom of Spain was just the latest in a long list of possessions gathered under the wing of the Habsburg family. The family could already claim a lengthy history at the centre of

European affairs, and on Charles's accession it stood at its greatest height. That this was so was due to the work of his immediate predecessor, Maximilian I. Through wars and, more particularly, marriage alliances, by the time of his death in 1519 Maximilian had amassed a sprawling empire. The Habsburgs reigned as lords of Bohemia, Hungary, Spain and the Americas, the Netherlands and assorted other principalities as well as the ancestral lands of Austria and the Tyrolean region. As Holy Roman Emperors they also commanded the dominant political power in Germany. Despite his success, throughout his life of chaotic energy Maximilian was faced with one emergency after the other. He waged almost continuous war with the French, and at one point was also locked in a war with powerful Venice. He had to cope with fractious German princes who constantly refused to follow his policy. All of which meant that he had continual need of soldiers and arms, not to mention the cannon that were the backbone of the armies of the time. To make matters worse, Maximilian was a thoroughly unreliable manager of the financial resources of his realms. As a result he had an insatiable and never-ending need for money. How could he raise it? His most obvious recourses – the levying of taxes and the sale of produce from his lands – were fraught with problems. Higher taxation often led to peasant riots, while the sale of produce yielded only a fraction of his needs. He was no doubt happy to accept the treasure from his colonies across the ocean. But his real salvation was to come from the liquation works of Saxony, Bohemia and the Tyrol.

To grasp how liquation came to underpin the ambitions of Maximilian it is necessary to understand some of the legal framework of the time. The laws pertaining to mineral exploitation were first formulated in a series of mining charters that were the product of the great days of the central European silver discoveries of centuries before. According to these charters, the lord of the land possessed a 'Regalian right'[18] to a share of the production from any mining activity that might occur within his lands. With that right came the responsibility to provide for the proper administration of the mines and smelters. Putting it another way: the lord of the land could claim a share of the output of the mines within his domains, and he could appoint whomever he saw fit to run them. And this is exactly what Maximilian did. His need for money, however, stopped him from waiting for the metal to be produced before he claimed it. Instead, he got into the habit of claiming it in advance in the form of huge loans. These he repaid by allowing his Regalian share of metal to be claimed by his creditors. The outcome was that many of the mines of the Habsburg lands came to be managed by Maximilian's bankers. Through this means families such as the Welser, Imhoff and Hochstetter rose to prominence.[19] But none was more important than the Fugger.

Their story begins with Hans Fugger, a weaver by trade, who settled in Augsburg in the year 1367. He left a modest fortune to his sons Andreas and Jakob. Jakob married the daughter of the master of the Augsburg mint. This man, Basinger, fell into trouble with his debts, having to be bailed out by Jakob. Soon after, Basinger opened a mint in the Tyrolean town of Hall, at the centre of the fast-growing silver district. In his gratitude to Jakob, Basinger invited him to participate in the buying of the silver needed for the mint. It was the beginning of a long and profitable Fugger association with the Tyrol.[20] Jakob Fugger died a prosperous man, leaving behind him five sons. The three eldest were designated to carry on the family business; the remaining two – Marcus and Jakob – were earmarked for the Church. When one of their number died, the two remaining merchant brothers asked Jakob to abandon the clergy and join them in the family business. This was ever after considered by the family to be a great stroke of luck, as it was Jakob, a man possessed of commercial genius, who pioneered the family's move into the two growth industries of moneylending and metals. In the decades that followed, Jakob the Elder's three merchant sons were to make the name of Fugger synonymous with vast wealth.

Jakob the Younger's first significant step was made in 1487. Archduke Sigismund of Tyrol was perennially short of money, and Jakob ventured to loan him a substantial sum. As security, Sigismund granted him a mortgage over the mines of Schwaz, and then over the entire silver production of the Tyrol. If the loans were not repaid on time the Fugger were to receive the Tyrolean silver production due to the Archduke. The following year Jakob and his brothers advanced a much greater loan to Sigismund. Under the terms of this loan the entire production of the Schwaz was to be delivered to the Fugger at a low price until the debt was paid. Then Sigismund handed over the reins of government in the Tyrol to his cousin Maximilian. The latter lost no time in further pressing the Fugger for more loans, thereby extending their hold on Tyrolean silver. In this way, all within a few years, the Fugger had come close to controlling the output of one of the most important silver regions in Europe. The family was gaining confidence both in their judgement as lenders and in their ability to successfully manage metallurgical concerns.

When extensive deposits of copper were found at Neusohl, in the distant plains of central Hungary, the Fugger set their sights on it. They faced formidable obstacles. For one thing the Hungarian nobility had little but their pride, and anti-German feeling within the country was high. In addition there were political and bureaucratic difficulties of all sorts for a foreign investor seeking to establish himself. Undeterred, Jakob and his brothers formed a partnership with the now very wealthy and influential Jan Thurzo.[21] Being of Polish extraction, Thurzo was able to navigate the political waters of the Hungarian court until he had secured for the Fugger and himself a lease over the entire district. Development capital was poured in. A *Saigerhütten* was built. The area around Neusohl was lacking in the swift-flowing rivers that were indispensable for driving the furnace bellows and the machinery for pumping and crushing. No matter. Thurzo, with the backing of the king, proceeded to construct large dams in the mountains some fifteen miles distant. The Fugger–Thurzo company was even permitted to export the silver directly instead of handing it over to the royal house. Neusohl soon was the greatest copper-producing concern in Europe. Some of the produce was sent to the Fugger-owned *Saigerhütten* in Villach. So important did this latter works become in the Fugger metals empire, they built themselves a castle there. It was named the Fuggerau.

There now existed four main copper districts – Mansfeld in the Harz mountains, the silver-rich copper mines in the Saxony-Bohemia region, the Tyrol and, last, Neusohl. Of these, the Fugger–Thurzo combine controlled Neusohl and a large part of the Tyrol. They were important suppliers to Spain and Portugal, establishing in Lisbon a warehouse to supply the India trade. A large part of the German (non-Fugger) copper was used in the brass manufactories of Liège, Aachen and Nuremberg. For several decades the European copper market was divided into Fugger and anti-Fugger camps.[22] Jakob and Thurzo tried to break into the brass market, but were repelled. Neither could they consolidate their hold on the Tyrol. Their enemies struck back, scoring a notable success when they convinced King Ludwig of Hungary that the family had been sending less than pure silver to his mints. Ludwig confiscated the Fugger property and jailed their representatives. We are entitled to wonder whether Ludwig was truly convinced of their guilt, for, having made his point, he restored their property a year later. Periodically the struggle would be resolved by a market share agreement. One such came into being in the early years of the 1500s. It soon failed, owing, it is said, to the Fuggers breaking the rules. The very possibility of such agreements demonstrates how organized the main mining districts were becoming.

They were also becoming organized in another way. The early decades of the mining Renaissance had an aspect similar to the silver rushes of the medieval era. Central Europe was again the land of opportunity, where with luck and hard work even the poorest adventurer could make his fortune. Martin Luther's father, the son of a peasant, became first a miner in the Mansfeld district, then a mine investor and finally a town councillor of Mansfeld, a position of considerable distinction in those days when a city government was an important political entity. Gradually these opportunities reduced as the industry matured. As the initial euphoria of economic recovery died down, the emphasis shifted from the opening of new mines to the better management of the existing operations so that investors could reap the rewards of their risktaking. And risktaking it was. As one observer put it: 'if we observe the matter rightly, people have put much more money into that Schneeberg and the hills around than they have taken out in profit'.[23] Once their investment was committed, mine owners were understandably anxious to focus on costs and production, imposing tighter controls and severer conditions on their workforce. They soon found that they did not have a completely free hand.

Following the medieval tradition of craft guilds, the mine workers tended to band together to form associations known as Knappschaft. The Knappschaft had a range of purposes – they provided a means for collective bargaining, served as a way of keeping out unwanted cheap labour and provided a communal safety net whereby some support could be offered to the widows of men killed underground.[24] They grew powerful, partly because the lords were inclined to indulge them and partly because the ownership of any single mine was frequently split into so many individual holdings that the investors were completely unable to act in concert. They became dependent on their hired managers. These men were kept busy with frequent conflicts over pay and conditions, and in their dealings with the Knappschaft they were by no means assured of victory. The Knappschaft were powerful enough to turn to a higher authority when the need arose. Archduke Sigismund's judicial officers in the Tyrol often met with delegations who came to complain of slow payment of wages, extortionate prices for food being charged by 'company stores' and unreasonably long hours of work. It was not at all uncommon for the judicial arbitrators to adopt a conciliatory approach. The workers received many of their demands, such as the right to the payment, at the shaft entrance, of a weekly wage, the banning of company stores, and an increase in the number of paid holidays allowed in the Tyrol. Following a march on Innsbruck of strikers from the Schwaz district, the underground shift was limited to eight hours a day, and workers were permitted to undertake private prospecting in shafts abandoned by the companies.

Such an imbalance of power (in hands other than their own) could not long be tolerated by powerful lords dependent for their very survival on the revenues of their mines. The Duke of Saxony provides a good example of this dependence. Fully two-thirds of his state revenues were derived from his Regalian rights over mining within his territories. It was only a matter of time before the official mood changed. The turning-point came when Maximilian I assumed control of the government of Tyrol from his cousin. His arrival coincided with Jakob Fugger and his brothers being granted the rights to all Tyrolean silver. So it was that these two powerful and ruthless figures replaced the benign rule of Sigismund and his reasonable officials. The impact was quickly felt. It began with a strike at Gossensass. The workers had the usual grievances over the conditions of shiftwork, pay and holidays, but it seems the tension had quickly escalated. Violence had been threatened, and the management and foremen had dispersed. The workers were demanding the right to take over the shafts.

They were invited by imperial officials to state their grievances in a mass meeting. As the meeting proceeded, Maximilian's troops entered the hall and arrested 27 of the strike leaders, and the rest of the meeting was devoted to a lecture to the remaining men, informing them that the 'Roman King is not satisfied but is extremely angered by recent events'.[25] The audience was then told that those who chose to continue the strike were to be banned from working in any mine within the Habsburg territories. Such was to be the tenor of industrial relations under Maximilian and his officers. Yet the miners remained a favoured group, as is demonstrated by the relatively lenient treatment meted out to them during the widespread popular revolts of 1525.[26] In any case the change to a harsher policy did little to slow the rate of growth in silver and copper production, nor of iron, tin or lead, all of which were found in quantities throughout central Europe. Advances in shaft ventilation, machines for hauling ore up the shafts and, above all, for removing water out of the mines kept pace with the growth.[27]

THE DAWN OF THE AGE OF AMERICAN SILVER

It is probable that the Habsburg Charles V looked at the experienced and skilful master miners of his central European domains and wondered how much better than the Spanish they would be at exploiting the riches of the Americas. Yet he could not simply order a mass emigration of Germans into the Spanish colonies. The colonists were a prickly bunch, proud and possessive of their territories, and Charles had to move carefully. After a while, however, it seemed that he had found a solution. Persistent rumours held that gold there was aplenty in the dense forests of what is now Venezuela and Colombia. It was forbidding territory, fiercely defended by its indigenous inhabitants, and no Spanish settlements had so far been established inland. The region was effectively vacant of Europeans.

Charles claimed it for himself, and then convinced one of the more important of his creditors, the Welser family of Augsburg, to loan him money in return for the rights to a large tract of the unmapped wilderness.[28] He granted them virtually unlimited powers over the area, and relaxed the restrictions on importing non-Spanish Europeans to the New World. As Charles no doubt hoped they would, the Welser chose Germans over the Spanish as their prospecting experts. Unfortunately, the rumoured gold proved elusive and the venture was soon in trouble. In the meantime, the Germans could not cope with the unhealthy climate and died in their hundreds. To add to their troubles, they excited resentment among the Spanish, who in turn were prepared to resort to murder to make their feelings known. All this was too much for a family whose business was moneylending and mine financing. Despite ten years of battling through the dense and disease-ridden forests, they had found little. In 1540 the Welser withdrew, leaving the field to hardier adventurers.

In the midst of the rush for gold, small but persistent quantities of silver kept appearing. The first silver ore to be worked by the Spanish in the New World is thought to have been at Tamazula, a region on Mexico's west coast.[29] It was a slow start, but it awakened the Spanish to the prospect that not only gold but silver too could be their quick path to riches. Next came the discovery of silver on an escarpment to the south of Mexico City. Five separate centres were soon being worked there, and collectively they came to be known as the Provincia de la Plata (the Silver Province). The area became a magnet for fortune seekers of all backgrounds. Leading figures in the colonial administration were among the most prominent investors. Cortes himself owned outright or held shares in 32 separate mines. Yet winning silver was quite a different matter to the digging and washing of gold from the alluvial deposits of the coastal plains. The alluvial gold lay free in the sandy soil and needed no hard digging. It also occurred naturally as a pure metal, requiring only to be washed from the sand. The silver ore, in contrast, had to be chipped out of the hard rock and smelted many times before it released its treasure. The Spanish could not even properly manage the exploitation of the sandy gold deposits. They were totally out of their league when it came to the skills needed for the operation of a profitable silver mine. Appeals home were soon being made, and again Charles authorized the dispatch of experienced German miners. A large group of them settled in Sultepec, where they established the smelting and refining practices that sustained the early industry. Water-driven mills were built for driving bellows and crushing ore, and the town was to remain a centre of American silver refining technology for many years.

The German presence in the Provincia de la Plata only serves to draw attention to the fact that the next phase of discovery and mine development in the New World seems to have proceeded without them. Certainly it is difficult to find many Germans who played a part in the series of silver strikes that occurred over the next couple of decades. The first of these was made by a small group of Spanish and Indians who left the northern town of Guadalajara in the summer of 1546. Their leader was Juan de Tolosa, and the party headed north-east with the general intention of prospecting for precious metals. As the party camped on one of the numerous rocky hills that punctuated that part of the central Mexican plateau, Tolosa made some friendly approaches to a group of Zacatecas Indians who, from some distance away, were watching the newcomers.[30] He no doubt was well aware that many of the mines to the south had been made known to the Spanish by local Indians appreciative of small acts of kindness from the feared foreigners.

His policy paid off. The Indians led the party to some stones that proved to be rich in silver. The find attracted a small number of hopefuls but seemed to show no particular promise. In this rather low-key way were developed the first mines of Zacatecas. Two years later was uncovered a vastly richer vein of ore and a rush of fortune seekers converged on the area. Then followed the discoveries that were to form, along with Zacatecas, the backbone of Mexican silver mining for the next two centuries. Guanajuato was founded in 1550, followed by Pachuco. Sombrerete came six years after that, and another nine years on Santa Barbara, much farther north, was established. These were just the main districts, and together with a myriad smaller finds they ensured that the centre of gravity of silver mining in New Spain swung well and truly away from the declining workings of the southern escarpment towards the rocky hills of the central plateau, hundreds of miles north of Mexico City.

A similar train of events was unfolding in Peru. Here the Spanish benefited from the many abandoned Inca workings. One of these was at Porco in the Andes mountains. Gold was found not far from the Inca capital of Cuzco. They were minor affairs, and still the restless Spanish probed and prodded the countryside. In April 1545, high up in the Andes, only fifteen miles from Porco, an Indian named Diego Gualpa climbed a distinctively-shaped conical peak in search of a rumoured Indian shrine. Such shrines frequently contained some gold or silver relics suitable for plunder. The peak was located at an altitude of more than 14,000 feet. Windswept and cold, it was surrounded by a barren, treeless expanse where not even enough grass grew to raise cattle. The story goes that Gualpa was thrown to the ground by a high wind. After scrabbling for a handhold, when he felt secure enough to observe his surrounds he found that he was gripping rich silver ore. He reported the find to various Spanish mineowners in nearby Porco, eventually persuading Diego de Villaroel to accompany him back to the site. Villaroel's inspection confirmed an abundance of silver lying very close to the surface. In this way was discovered the silver mountain of Potosí.[31]

Indians and their families, as well as Spanish, flocked to the forbidding peak once its great wealth became apparent, and within four years it was producing twice as much silver as all of New Spain. The makeshift camp was full to bursting. One estimate put its population at over 100,000. This would have made its inhabitants more numerous than contemporary London and, in European terms, fewer only than Paris, Naples and Venice. Surely it was the biggest mining boom town in history. As the first flood of its precious metal flowed into his treasury,

Charles V was moved to grant the sprawling settlement the title of Imperial City, with a coat of arms bearing the legend 'I am rich Potosí, treasure of the world, the king of all mountains and the envy of all kings'.[32] Germans were absent from the scene; instead, the local Indians provided the technical know-how for treating the difficult ores and a system grew up whereby the Spanish owned the mines and the Indians the smelters. Both grew rich. So, too, did the merchants from lower down the mountainsides. Everything from food and clothing to mining tools and lead for the smelters had to be hauled up the steep trails. The New World was becoming a silver country. When the silver from Mexico and Peru was added together, by the 1550s the total equalled, or perhaps even exceeded, the quantity being produced in all of central Europe.[33] But, for the moment at least, the German industry reigned supreme. Its magnificence was well-captured by its chronicler, Georgius Agricola.

GEORGIUS AGRICOLA AND HIS TIMES

Georg Bauer (the surname means 'peasant') is distinguished in the history of metallurgy as one of its first and most influential writers and commentators. He was born the year after Martin Luther in the Saxon town of Glauchau. Graduating from the University of Leipzig in 1518 he entered the teaching profession, securing the post of vice-principal in a school in the city of Zwickau, where he taught Greek and Latin. Moving next to the University of Leipzig, he then headed south to Italy to taste the fruits of the new learning. In Italy he acquired an interest in the sciences. Bauer must already have been a scholar of some note, as it was during this time he made the acquaintance of the famous humanist Erasmus, a friendship he was to maintain until Erasmus's death in 1536.[34] By this time Bauer had assumed the name Agricola (Latin for 'peasant'), it being the custom of those times for teachers to Latinize the names of their pupils. He returned to Germany to take up a post as physician in wealthy Joachimstal. From there he began to concentrate on the metals industry on his doorstep.

In a world of uncertainty and ruthless men vying for power, here, at least, must have appeared something securely prosperous. Adding to its attraction was the fact that despite the growing dominance of the richer capitalists, there was still plenty of room for small fortunes to be made by newcomers. Agricola himself invested in the Gotsgaab (God's Gift) mine. The dividends from this helped sustain him and his family for the rest of his life. Once settled in Joachimstal, Agricola began to spend his time travelling the mining districts, observing the activity and writing a series of books on mineral subjects. He published a book of four essays, the third of which, 'De Natura Fossilium' (On the Nature of Minerals), is celebrated as the first systematic attempt to classify minerals according to their properties. He began work in 1533 on what was to be his masterpiece, *De Re Metallica* (On Metals). From the start it was an ambitious work intended for a Europe-wide readership. The manuscript was not finished until 1550, and not published until 1556, one year after his death. It was a comprehensive treatment of all aspects of mining and metallurgy, ranging from how best to search for and identify mineral outcrops to the most effective ways of smelting difficult ores. Written in a strikingly easy-to-read and matter-of-fact style, *De Re Metallica* remained for 200 years the standard reference work on its subject.

When Agricola began his masterpiece the metals industry was taking on a majestic aspect. It enjoyed the interested patronage of the greatest lords and attracted the most ambitious merchants. It was a truly international scene, as is shown by the occasion when the sovereign of the Christian state of Muscovy, Ivan the Terrible, requested of the King of Hungary the services of mining experts to assist his subjects in locating and working metal ores. On being rebuffed, the Tsar extended his enquiries to Germany, from where Hans Stitte brought a troop of iron workers. Other Germans helped establish silver and copper works near Arkhangel. From across the English Channel came cheap lead stripped from the roofs of Catholic monasteries during the tumultuous years in which Henry VIII was creating his English Church.[35] Some was exported via Seville to New Spain, where it was used in the silver smelters of the Provincia de la Plata and Zacatecas. Most of the rest found its way into the

Saigerhütten of Germany.

The Fugger and others like them had grown fabulously wealthy. By the time of his death Jakob was known as 'the Rich'. His son Anton took the fortunes of the firm to even greater heights. This was despite the loss of Neusohl. This calamity had occurred after a new mine manager, who must have been a steadfast character, had convinced the Hungarian court that the Fugger were making immense profits while paying a price for the copper that barely justified its processing. The nobles terminated the Fugger contract. While perhaps not ever having cause to regret their decision, they did not greatly profit from it. They tried at first to sell the copper by themselves, but after a time realized that they needed the merchants who knew their way around the great cities of Europe and who could invest needed capital in the works. The man they chose as their new agent was Mathias Manlich. As rapacious as any of the Fugger family, after wresting control of the Mansfeld production out of the hands of the local nobles, he came close to a single-handed dominance of European copper sales.[36] Such was the nature of a business so closely connected with the ambitions of kings.

It was a risky business. One prominent banker to fall upon hard times was Ambrosius Hochstetter. After a good run of successes he had lost a fortune in his attempt to create a monopoly over the European mercury market.[37] Following that disaster, a spice ship lost at sea and a goods train lost to bandits were enough to ruin his business. As if that were not sufficient, his son, Joachim, and a nephew gambled away the remnants of the family fortune. They then absconded to England, leaving Ambrosius to face his creditors and end his days in prison. The disgrace seems not to have unduly affected the career of Joachim. The next we hear of him he is in the employ of the German firm of Haug, Langnauer & Company and visiting the court of England's Henry VIII, asking for permission to prospect for gold. Haug, Langnauer & Co. was a mining 'multinational' with offices in all the usual centres – Augsburg, the Tyrol, Hungary, Poland and Bohemia. They had already been engaged by the English king to help reform his debased silver currency. On the strength of the interview Hochstetter was appointed to be principal surveyor and master of the mines of the realm. It was a shortlived assignment, as he was soon on his way to Hungary, having been persuaded by the Ambassador of Hungary to assume the management of Neusohl.

The central European mines had enjoyed 50 years of rapid and virtually uninterrupted growth. Although this period was irrevocably over by the time of his death, Agricola would scarcely have noticed the slowdown. There would have been no diminution in the prosperity of the more successful mines, and the paucity of statistics meant he would have seen no statistical evidence of the overall decline that was then occurring. The beautiful illustrations provided in *De Re Metallica* show with great clarity the sophistication of the industry. Elegant and clearly powerful machines for hauling water out of deep pits, machines for raising ore and men up the shafts, and rows of stamps for crushing – all grace the pages of the book. Some were driven by water, others by horses and still others by men. One of the more curious devices was the rag-and-chain pump. This consisted of a series of hollowed-out logs that created a pipe extending from the shaft mouth down to the lowest point of the workings, where water was collecting. Through the pipe was passed a chain along which at intervals specially fashioned rag balls were fixed of a size that fitted snugly into the pipe. As the chain was hauled upwards the balls would trap the water in the pipe and push it up and out of the shaft. The whole arrangement was powered by a waterwheel.

Most of the larger mines were well capable of supporting the advanced array of equipment shown in Agricola's woodcuts, although there were certainly very many workings where such machines as these could not be afforded. And the old methods long persisted in some shafts where the machines no doubt could be afforded. One of the latter was Falkenstein in the Tyrol, which employed daily 600 men whose task it was to form a continuous human chain along which, day and night, buckets of water would be passed up and out of the underground workings.[38] When the management finally installed a water-powered rag-and-chain pump, its total cost amounted to only a year's wages for the bucketmen.

In the foreword to his book Agricola devotes space to a spirited defence of the industry, against what must have been a small army of critics. Mining and smelting, it seems, was being criticized as too dirty, too noisy, dangerous, environmentally damaging, financially ruinous to many and inherently an unstable boom and bust industry. For each of these accusations Agricola has a reply. It is hard to imagine that they posed any serious threat. On the contrary, so important to the princes were their metal revenues that the mining fields seem to have been run in some respects like military camps. Agricola tells us that the state-appointed inspector, the Bergmeister, was every day supposed to visit one or more of the shafts under his care and audit every set of books once a week.[39] In the course of his lengthy book Agricola makes little mention of the two real threats facing the industry – the destruction of war and the competition from new districts in faraway lands. Both became realities quite soon after his death. Indeed, during his lifetime the religious conflict that culminated in civil war was becoming heated. So much so that even so moderate and respected a man as Agricola became caught up in the struggles, and it is said that in his final years his former friends and colleagues shunned him for his refusal to abandon his Catholic faith. As for competition from distant lands, the decade after 1550 saw the first undeniable decline in the silver output of central Europe. Production was moving across the Atlantic. This shift received a huge boost from a German master miner who made his way to the Spanish port of Seville, carrying with him a new metallurgical process.

MERCURY ALMAGAMATION

At the beginning, the silver deposits of New Spain had been extremely rich. They had to be. The smelting was done using lead metal to dissolve the silver out of the surrounding ore on the same principle as that used in the mines of central Europe. This lead was expensive and, owing to the technical incompetence of the Spanish, was wastefully used. Wood for fuel was also a costly commodity, as the silver mines were rarely in heavily wooded regions. Add to this the costs of building the smelting equipment and of constructing any machinery, such as stamp mills for breaking down the pieces of ore, as well as the cost of hiring labour, and many mines began to struggle. Without very rich ore they simply could not be made to pay their way. Compounding these problems was the fact that the ore was becoming more difficult to smelt as the mines went deeper. Smelter-waste, rich in silver, was beginning to pile up throughout the Provincia de la Plata. This is not to mention the huge reserves of lower-grade ore that the miners did not think even worth digging.

In an attempt to remedy their worsening plight the miners used their political muscle to convince the King to halve the rate of gold taxation. They also resorted to virtually enslaving the Indian population, justifying themselves by claiming that they were acting for the good of the Spanish kingdom. The laws of 1542 that decreed an end to slavery were ignored. Then came the arrival of a new Viceroy, Don Luis de Velasco. Refusing to tolerate the widespread flouting of a royal decree, he set about enforcing the abolition of slavery. Overnight the costs of silver production leapt upwards. It is at this juncture that one of those revolutionary advances in metallurgical technology – like liquation a century earlier in Europe – changed the entire economics of the industry and in the process changed the history of the world.

The hero of the piece was an unlikely revolutionary, a 50-year-old cloth merchant named Bartolomé de Medina.[40] Medina had developed an interest in the smelting of silver. It was an unusual hobby for a cloth merchant. No doubt, living in Seville, he had plenty of exposure to the stories of the silver wealth of New Spain, and as a successful and shrewd merchant he may have suspected that there was plenty of room for improvement in the smelting of it. Whatever the case, Medina became something of a local authority on the smelting of silver ores. He attracted men skilled in the art, an attraction made easier by the fact that all miners returning from or on their way to the Americas had to pass through Seville. Sometime after 1550 he received a visit from a German, a man who has become obscured by the mists of time and who is known to history only as Master Lorenzo. Master Lorenzo brought with him an entirely new approach to the winning of silver from its ore. He had sought out Medina because he needed financial backing and a 'Spanish connection' to be able to travel to New Spain.

The process Lorenzo brought to Spain has come to be called mercury amalgamation. This technique, as applied to silver and gold metal, had been known and in use for decades. In *Pirotechnia* we can find, ten years before Lorenzo's journey to Seville, a thorough description of the process. Clear as it is, there remains doubt about whether Biringuccio's description applies to the amalgamation of ores as well as pure metal. The confusion has led many to question whether Lorenzo (or Medina) actually devised a new process or simply copied an existing one. Whatever the answer to this, the true significance of the Lorenzo–Medina partnership is that they took a little-appreciated metallurgical technique and applied it, with resounding success, to a vast untapped opportunity.

To test the process Medina and Lorenzo took silver ores from the old abandoned Rio Tinto deposits in southern Spain, ground it up as per Lorenzo's instructions, added strong brine and mercury, stirred and waited. The result after four to six weeks was a pasty metallic amalgam in which the mercury and silver had combined, leaving everything else to be drained away. The one step remaining was to boil off the mercury. Bright metallic silver was all that remained. It worked beautifully. The most impressive aspect was that it also worked on the low-grade ores and tailings that were considered to be waste under the smelting techniques of the time. Full of hope the pair prepared to sail for New Spain. It is here that Lorenzo vanishes into the mist. The Spanish authorities refused him permission to sail. The reason given was that he had not been resident for the required ten years in Spain. In the light of previous German migration this seems a dubious excuse, and probably reflected the hardening attitude of the Spanish towards the unwanted foreigners. So it was that in 1554, leaving his family behind, Bartolomé de Medina set sail for New Spain without his collaborator.

On his arrival in Mexico City he lost no time in heading for the mines. Pachuca, the town he selected as the site for his experiments, was a curious choice. It was one of the newest silver fields in New Spain, and being still in its infancy, its rich surface ores would presumably have meant that the need for his process was not very strongly felt. A more logical choice would seem to have been the older and declining districts of the Provincia de la Plata where, what is more, there was a large population of German miners experienced in the local conditions who could bring a more disciplined and insightful approach to the application of mercury amalgamation. Whatever his reasons, it was in Pachuca that Medina confidently undertook his demonstrations for the sceptical local miners. Disaster struck. Inexplicably, and despite his total confidence, the demonstrations were a complete failure. Instead of forming the coarse dull grey paste that showed that the amalgamation reaction had succeeded, the mercury remained unchanged in shiny droplets dispersed throughout the mixture of ground ore and brine. Unable to understand what he had done wrong, Medina repeated the exercise, failing each time. It was only after an extended period of feverish effort and 'mental anguish' – as Medina himself describes – that he managed to get the process working. The key to it appears to have been the addition of the right amount of a substance called 'magistral' (probably copper sulphate), a chemical that was naturally occurring in the Rio Tinto ores on which Medina had conducted his tests in Seville.

Despite his mental anguish, it is remarkable how quickly Medina was able to adapt his process to the local conditions. A single trial could take six weeks to perform, yet within eighteen months he had the process working to the point where he could justly apply for a patent. In the incomparably more scientifically sophisticated nineteenth and early twentieth centuries, it took longer for the flotation process finally to succeed and at least as long for the Bessemer steel process to be made to work. It seems reasonable to speculate that correspondence with his friend Lorenzo holds the key. Biringuccio had stated that the successful refining of metallic silver required 'vinegar, verdigris and corrosive sublimate'. It does not seem out of the question that Lorenzo – who would have been familiar with Biringuccio and must have been a very competent metallurgist – advised his friend on the range of additives he might usefully try. Whatever was the case, the amalgamation process was ready for patenting by the early months of 1556. It was the same year that *De Re*

Metallica first appeared in print. In Agricola's famous and comprehensive work there was no hint of the new process.

Amalgamation became known as the patio process. The word derives from the use of large stone-paved open-air baths that contained the mixture of ground ore, brine, mercury and magistral. These baths were originally rectangular in shape and gave the appearance of being terraces or *patios*. The first *patios* full of ground ore and water were mixed by the action of several Indians sloshing and shovelling through the thick poisonous mud, one of whom liberally squirted mercury over the mixture from a perforated sack. Later on the baths were to assume a circular shape to better allow the mixing to be conducted by a mule-drawn stirring device. Before the process could work effectively on a large-scale, some mechanical means had to be found to grind the ore into a powder sufficiently fine. None such existed in the New World, and here again a German, one Gaspar Loman, steps in.

Loman was living in Sultepec when word arrived of Medina's wonderful new process. His first reaction was to obtain a licence to prospect for mercury-bearing ores. When it became apparent that Medina possessed no efficient means of grinding large amounts of ore, finding such a means became the German's next project. In the event he simply copied a machine that had long been in use in his home country for the grinding and amalgamation of fine-grained gold-bearing rock. This so-called *ingenio*, illustrated in the pages of *De Re Metallica*, consisted of a water-powered grindstone and a series of water-powered paddles to stir the mix. Loman seems to have been quite a salesman, because his next successful step was to convince Medina to agree to a joint patent covering both the patio process and the *ingenio*. In the event, the machine never caught on. Water was scarce on the central plateau, and the preferred means of grinding became the *arrastra*, a mule-powered heavy rock wheel that, when rolled over the ore, crushed it into fine sand.[41] The Medina–Loman partnership was nevertheless a fitting one, for in this combination of fine grinding and chemical action had been found the metallurgical key to the silver riches of the New World.

All that remained was to secure a sufficient supply of mercury. No amount of searching was ever to reveal a local source anywhere near large enough to supply New Spain, and soon attention turned to the mercury mines of Europe. Of these, there were only two of any size. By far the largest were the Almadén mines located in Spain itself, and which Charles V had already leased to the Fugger family.[42] With the invention of the amalgamation process, the sleepy township of Almadén now assumed a vital importance for Spain. So much so that if Almadén had perchance been located in a different kingdom, the history of Europe may have taken a different course. As it was, the Spanish crown, after some hesitation, imposed a monopoly on the sale of mercury in the Americas and controlled its export through the port of Seville. The monopoly met with determined protest. The Viceroy of New Spain reported a petition that described the imposition of the monopoly as akin to 'placing a monopoly on bread or meat, for it is understood that the sustenance of the land depends on the mines of silver and they cannot be sustained without mercury'.[43]

Mercury amalgamation had important consequences for the structure of the American silver industry. The first was that it resulted in a concentration of ownership and of wealth. The amalgamation process required a large investment before an operation could be profitably run. The initial requirement was the necessary machinery for grinding the ore. Without it, the manual grinding of the low-grade ore, even on a small scale, was too arduous a task to be worthwhile. It also needed a patio of a reasonable size before the expense of the equipment required for the collection and distillation of the amalgam could be justified. The small man was squeezed out. A second consequence of the amalgamation process was the extent to which it permitted government control and taxation of the industry. With the small-scale smelting operations, tax evasion was relatively easy. With the advent of amalgamation, the average size of an operation increased and their number decreased. Thus effective control became easier, as did the collection of the royal fifth. Mercury taxation revenues were especially easy to collect. Little wonder that successive sovereigns adhered stubbornly to their mercury monopoly in the face of vociferous calls over many hundreds of years for its abolition. Many argued that the policy was an actual hindrance to the production of silver. This may indeed have been so, but to the Spanish crown that was not the point.

THE ARCHITECT OF COLONIAL PERU

It took another couple of decades before Potosí underwent the same decline as had been experienced in New Spain. Once again, the silver was certainly there, but even the skilfully operated Indian furnaces could not coax it free from the rock. As it had in New Spain, mounds of smelter waste began to pile up, rich in inaccessible silver. By 1570 silver production had halved from its level of fifteen years earlier, and Potosí was in full decline.[44] Making matters worse, there was worsening conflict throughout the Vice-Royalty between Church and landowners over the correct treatment of the Indians. Then there was the issue of the Indians themselves. Despite the best efforts of missionaries and educators, they refused to adopt the lifestyle and habits of good Spanish Catholics. The Habsburg Philip II, successor to Charles, convened a council to consider how best to deal with the many problems. One of the councillors was Francisco de Toledo, and to him fell the task of carrying out the council's recommendations. Chief among them was the rescue of Potosí. Philip had made that quite clear. With typical attention to detail, the King further directed Toledo to establish royal ownership over the silver-rich waste, as these would likely be readily susceptible to amalgamation.[45]

Toledo was a cold and distant man of a type not unknown among his Spanish contemporaries – an idealistic reformer, a man of considerable intellect and insight, with the Spanish passion for order and administrative thoroughness. He arrived in Peru in the year 1570. Sometime after his arrival he summoned a special council to consider the question of Indian labour in the mines. After due deliberation, the council concluded unanimously that mining was in the public interest and, as a result, the use of the forced labour of Indians was justified if this was the only way in which an adequate workforce could be assembled. This was not, as might be thought, necessarily a cynical or self-serving conclusion. On the council were several men of the Church with long records of criticizing abuses perpetrated against the Indians. Toledo himself was indignant, even outraged, at the actions of the landowners in exploiting the Indians, and he seems to have held a genuine concern for their welfare. The council's conclusion was determined by the fact that Indians would not work in the mines by choice, and the mines were believed to be essential to the economy of both the colonial society and of the Spanish kingdom. Acting, therefore, with the support of King and council, Toledo gave instructions for the establishment of the *mita*, Peru's forced labour system.[46] The system stipulated that sixteen of the highland provinces of Peru were to provide at any one time one-seventh of their eligible males – those aged between eighteen and fifty – to work at Potosí. These drafted labourers were to be paid a reasonable wage, and it is probable that Toledo foresaw a system in which the Indians would benefit from their sojourn by refining their own silver and accruing some small profit to take back to their homes.

The policy was, in Toledo's mind, designed only to ensure that sufficient labour would be available for the mines. The selection of the word *mita* is instructive of his intention. The word derives from the Inca system of labour tribute. Toledo no doubt felt there were some parallels with his system and the communal projects that benefited all under the relatively benign rule of the Inca. He was tragically mistaken. From the beginning the *mita* was to become a ravenous beast, devouring the able-bodied men of Peru. These unfortunates had none of the rights possessed by the free Indians in Potosí, and they were assigned the tasks that no others would take. Least welcome of all was the job of carrying baskets of ore from the bottom of the shaft up to the open air. It was murderous work. The basket when fully loaded often came to more than twice a man's body weight. The shafts were now sometimes 100 yards or more in depth and the ladder a series of slippery logs with handholds cut into them, or perhaps two parallel logs with leather straps slung between them. On at last reaching the top, the exhausted and sweat-soaked *mitayo* would be met by the freezing wind and often blows from his Indian or Spanish supervisor.[47] It was not unknown for him to be sent back down because he had brought insufficient ore to the surface. So feared was the call to the mines that it became a custom among the Indians to say a requiem mass for those beginning the journey to the great silver mountain.

The need for a local source of mercury was particularly acute in Peru because of the distances involved. It cost five times as much to transport mercury from Almadén to Potosí as from Almadén to New Spain. So when Enrique Garcés noticed the use of a vermilion paint among the local Indians, it set him thinking. Vermilion was a dye that in Europe was made from mercury. He set about looking for the source of the dye, and soon had located a good-size mercury deposit. A second one, larger and richer, was found four years later near the trail between Lima and Potosí. It was named Huancavelica, and soon its production was so prolific that Peru was freed from dependence on faraway Spain for the liquid metal that was to prove its lifeblood. Huancavelica grew to be a mining enterprise on a scale second only to Potosí in all of Peru. It was soon notorious. Worst of all was its richest mine, the Cerro Rico. Here the workers entered each day through a magnificent sculptured opening in the side of the mountain, above which was carved the royal coat of arms. Once through the grand entrance, conditions deteriorated until the worker found himself in a maze of narrow, airless paths that wound their way toward the mercury-bearing ores. Crowded into these tiny workplaces, the miners would labour in the dark enveloped in the choking mercury and arsenic fumes that wafted off the rockface. Visitors, presumably familiar with other mining environments, were horrified by the smell and noise. One visitor wrote that for the Indian labourers 'sending them to such work is sending them to die'.[48] The carnage was such that later conscience-stricken Viceroys would try all sorts of policies in order to reduce Peru's dependence on the mercury of Huancavelica.

Having settled the principle of Indian labour, Toledo departed his capital for a tour of the lands in his care. In Potosí he set about putting in place his plan to revive the failing fortunes of the town. He announced a reward for the first person who successfully used the patio process on Potosí ores. It was soon claimed. In the wake of amalgamation a great wave of confidence swept across the mountain. Investments in milling operations and *patios* continued until the little Indian furnaces had virtually disappeared. The Spanish came to dominate this new industrial landscape and, with the exception of one or two, the Indians were excluded. The *patios* – fed with the silver-bearing 'waste' rock that under the previous smelting technology had been unusable – had an explosive impact on production, and silver shipments from Peru soon overtook those of its northern neighbour.[49] In the process the ramshackle town was given a facelift. It is estimated that at its height 160,000 people were making Potosí their home. Its wealth and the carelessness with which it was spent was remarked on by all.

A GLOBAL SYSTEM OF TRADE

Spain drew few real benefits from its bonanza. Philip II, possibly suffering from an exaggerated sense of wealth, managed to spend his windfall with such abandon that he defaulted on his debts in 1557 and again eighteen years later, in the process sending the Fugger on their way to obscurity. His financial troubles had their root cause in his incessant waging of war. Like his father before him, Philip had taken on a virtually impossible political task. Bequeathed the Netherlands as part of his empire he spent his entire reign trying to cope with the rebellious people of that country. Protestant England and Catholic France judiciously offered assistance to the rebels – never enough to allow them full victory, but enough to keep the Spanish well and truly embroiled. In frustration Philip struck out at France, spending yet more money in fruitless campaigns.

Being the sovereign over the New World roused the jealousy of his neighbours. Where was it written in Adam's will, asked Francis I of France, that the New World should be reserved to the Spanish king? Jacques Cartier was dispatched by the French king to North America to search for gold and a northern route to Asia. As for the American treasure, it was considered fair game. As early as 1523 a French naval squadron led by Jean d'Ango had commandeered several of Cortez's treasure ships and sailed them to France along with their cargo of Aztec gold.

The sea captains of France, England and the Netherlands all gained experience in the waters of the Caribbean and began to prey on the Spanish settlements.

Privateering, as it was called, developed into a lucrative business, attracting wealthy investors willing to put up the money for the ships in return for a share of the loot. For the most part the privateers concentrated on soft targets, such as coastal communities, where they could loot and kill without encountering a determined resistance. One of the bolder forays resulted in the town of Havana being plundered and burned by French privateers. None was more daring than the English farmer's son, Francis Drake. Drake made his first fortune and his name by intercepting the mule train carrying Potosían silver and New Granadian gold into the Panamanian port of Nombre de Dios.[50] He and his men made off with a fortune in gold; the silver they left behind because it was too heavy to carry. Years later he set out again in his ship, *The Golden Hind*, rounding the southern tip of South America and making his way up the sleepy and unprotected west coast, plundering as he went. The knighthood bestowed on him on his return by Elizabeth I was one more provocation that led to the launch of the Spanish Armada.

Not even Drake dared to attack the ships bearing bullion across the Atlantic. These treasure fleets were most certainly not a soft target. For in response to their early losses, the Spanish had begun to organize annual convoys in which the yearly haul of precious metal was transported from the New World back to Spain under armed guard. These convoys operated as an *ad hoc* arrangement until the Crown ordered that all ships bound for Spain must sail in two armed convoys – one from New Spain, the Flota, and one from Panama, the Galeones. After loading the annual haul of bullion, each fleet would sail to Havana, where they were to link up for the trip back to Spain. Despite the precautions, the months of July and August must have been anxious ones for the Spanish monarch until the treasure fleet was sighted on the horizon. Once in Europe, the silver quickly found its way into circulation. Inflation, unknown for centuries, returned. Some have credited this inflation with causing an awakening (or re-awakening) of the entrepreneurial spirit that finds its expression in trade and ultimately in capital investment.[51] Indeed, one of the more positive impacts of the American treasure was the stimulus it gave to trade. In time it came to underpin a global system of maritime commerce.

The way was led by the Portuguese who followed in the wake of Vasco da Gama. Their first obstacle to establishing regular commerce with the Indies had been the Arabs who controlled the sea lanes of the Indian Ocean. In 1509 Portuguese sailors defeated them in the battle of Diu and control of the Indian Ocean was theirs. They would hold it for the next 100 years. Sailing east from their bases at Goa and Cochin they next captured the strategic town of Malucca. This gave them control over the gateway to the rich spice- and gold-producing islands of the Malayan and Indonesian archipelagos. They then proceeded to build up a complex Asian trade. The general schema of the trade was that the silver of the Americas, European copper and mercury, coral and a small number of manufactured items were packed into the holds of Portuguese ships bound for Asia. Some African gold was picked up along the way. The first stop was the eastern ports of India. Here the metals – especially the copper – would be exchanged for fabrics, which were then carried to Malucca and bartered for nutmeg, mace and gold. In the early days a ship's captain would then have turned his little vessel back, stopping in Ceylon and southern India to trade the rest of his precious metals for cinnamon and pepper before setting sail for home.[52] The Portuguese were astute businessmen, and on this general formula there were many variations. Of course, to assert their authority in an utterly foreign and hostile world they had to be more than astute, and a good part of their success was owed to the fact that they were also unscrupulous thugs who, armed with muskets and sheathed in armour, could fend off the depredations of pirates and saw no harm in bringing that same force to bear on local princes. In the 1540s Portuguese trading bases were established in two new places. The first was at Ning-po in southern China; the second, a year later, at Nagasaki on the southern tip of Japan. The timing could not have been better.

China's Ming dynasty had by then been in power for nearly 200 years. The first emperor, Hung Wu, had come to the throne after widespread revolts had succeeded in toppling the last drunken Mongol khan. The inflation and resentment occasioned by the Mongol system of paper money are considered to have been two of the principal hardships that emboldened the populace into armed insurrection, and, understandably, Hung Wu decided early in his reign to reform the currency. He proceeded in a curiously half-hearted fashion. Copper coins were minted, but neither Hung Wu nor his successors could bring themselves to embrace silver money.[53] The merchant class, on the other hand, regarded silver as the most reliable of currencies. The Imperial policy governing the mining of silver was torn between these conflicting views. In 1435 a decree ordered the closure of all silver mines. It was followed by a supplementary decree that pronounced the death penalty for all persons found secretly smelting the metal. In 1442, at the prompting of court officials, some of the mines were re-opened. They were closed again eight years later. Some years further on we read of prisoners being sent to work the silver of Yunnan Province.[54] Another policy reversal brought this to a halt shortly after. While things went on in this fashion, the merchants turned increasingly to using silver bars – coins being unobtainable – for their larger transactions. This seems ultimately to have had its effect on the Imperial court. In time it became permissible to pay taxes in silver, although the limited supply of the metal was such that this was only possible in certain regions. The southern coast of China was one such region, for the merchants there had begun trading their silks for the silver of Japan.

The emergence of Japan as an important metal-producing region followed a path similar to Renaissance Europe. The stimulus to the production of precious metals stemmed from the civil wars that tore the country for nearly 100 years. Warring princes needed money to pay troops and to reward their faithful followers, and the production of gold and silver, as well as of iron and copper, was fostered and protected. Production grew, and eventually the precious metals found their way into ships bound for China. It was a dangerous trade. This was in part because of the pirates who terrorized the straits and in part because the Chinese emperor frowned on contacts with the vigorous and increasingly independent Japanese. A series of particularly large silver strikes, including Iwami at the southern end of Honshu and Tsurushi on the island of Sado off the west coast of Honshu, were made just about the time when the Portuguese set up their trading post in Nagasaki. Ten years later the Imperial court in Peking pronounced the logical conclusion of its erratic monetary policy: all taxes were now to be paid in silver. It was known, presumably with black humour, as the 'single lash of the whip'.[55] The new policy created a vast and insatiable demand for silver, and it came pouring in.

Under the new stimulus the production of Japanese silver grew in leaps and bounds. It was aided by imported Chinese technology and given a further boost by the introduction of the liquation process, given to Sumitomo Jusai by an unknown European.[56] A series of strong leaders brought the civil wars to an end and unified the country. The first was Oda Nobunaga. He was succeeded after his death in 1582 by one of his generals, Toyotomi Hideyoshi. Next came Tokugawa Ieyasu.[57] Each recognized the importance to a unified empire of a common currency and buoyant trade, and so the fostering of metal production became an important aim of the state. Gold and silver became so plentiful that one chronicler was able to write: 'Ever since the advent of Hideyoshi, gold and silver have gushed forth from the mountains and from the plains in the land of Japan ... in the old days no-one as much as laid an eye on gold. But in this age there are none even among peasants and rustics, no matter how humble, who have not handled gold and silver aplenty.'[58] The Portuguese were referring to Japan when they wrote of the 'silver islands', and it did not take long for these bearded 'southern barbarians' to secure a dominant position in the trade of Chinese silks for Japanese silver. It was a highly profitable business.[59] Even more lucrative for the enterprising Portuguese was the silver they carried to Chinese ports from Lisbon, metal that had originated in the *patios* of Peru and Mexico.

The phenomenal success of Portugal excited envy throughout Europe. Not least envious would have been the Spanish, who saw the fruits of their empire being turned to spectacular profit by their nearest neighbours. Unsurprisingly, they wanted some of the action and eventually managed to secure a defensible trading post near China. Manila was located inside a magnificent bay on an island that formed part of the Philippines, an island chain named by Magellan in honour of the Habsburg King of Spain. Once established, the Spanish commenced a vigorous cross-Pacific trade. Fleets of so-called Manila Galleons would leave the Mexican port of Acapulco laden with silver while Chinese merchants departed from the south China ports in junks laden with silks, delicate blue-and-white Ming porcelain and specially-crafted furniture and other goods designed to please the Spanish.[60] In later years, Chinese mercury also found a place in the holds of the junks. With the Bay of Manila as a backdrop, Spanish and Chinese would meet to barter and bargain. Manila became a major city. Many of the Chinese merchants settled there, their numbers rising to tens of thousands by the turn of the century.[61]

The Manila route grew in importance relative to the Japan and the Cape of Good Hope routes. Indeed, it may have been the most important of them all. It is difficult to know this for certain, however, because most of the trade was illegal. It never had the sanction of the Spanish monarch. The edict of Ferdinand still applied: all bullion leaving the Americas was to be sent via Seville, while all goods bound for the Americas were to be shipped from Seville. Powerful interests in Spain with access to the ear of the King were not about to let this valuable monopoly be ignored. Yet neither were others about to pass up their chance for riches, and Acapulco became the smuggling capital of the Americas. Silken robes today comprise the national dress of Mexico, mute testament to the great days of the Manila Galleons. So it was that over time an organized system of trade developed that linked together ports throughout the Atlantic, Indian and Pacific oceans. Silver, silks and spices were the oil in the wheels of this global commerce.

A PLATEAU IS REACHED

American silver seriously weakened the mines of central Europe. They were already becoming expensive to run, both because of their increasing depth and the diminishing richness of their ore. Now the inflation caused by the flood from the

Spanish colonies meant that a given quantity of silver could buy fewer goods. Silver had underpinned much of the prosperity of Agricola's industry, and the copper produced from the liquation works was sometimes profitable only because of the silver that accompanied it. Now as the silver lost its value, it was brought home that the copper which came with it was expensive. The silver–lead mines were in a similar situation. It was enough to push many of the poorer operations into bankruptcy. Others fell victim to the civil wars that ravaged central Europe, triggered by a combination of nationalist rebellion against the Habsburgs and religious tensions between the Catholics (the faith of the Habsburgs) and reformist sects such as that formed by the followers of Martin Luther. Civil strife in Germany caused a steep decline in demand for the brass goods that had for long been the mainstay of Mansfeld and Bohemia. Meanwhile, the Spanish Habsburg occupation of the Netherlands devastated that country and virtually put a halt to trading activity in the Dutch ports. As if that were not enough, a war between Denmark and Sweden closed to traffic the straits separating the Baltic Sea from the Atlantic. The resulting disruption to the copper trade plunged Neusohl into such disarray that a skeleton staff of 150 workers was all that remained, and these only to work the pumps that prevented flooding of the underground galleries.[62]

Now came the chance of the deposits whose copper and lead content were high enough and whose costs were low enough to justify working them for those metals alone. They were especially favoured if they were located in lands relatively free of civil strife. In this way the centre of gravity of metal mining and smelting moved from central Europe to various regions on the Continent's periphery. In England the low-cost and silver-free lead deposits in the north and west overtook the old faithful fields of Tarnowitz and Bleiberg. Iron-making migrated to England and Sweden. And a rich copper deposit in northern Sweden came to prominence.

The tale of Sweden's Stora Kopparberg – the Great Copper Mountain – begins perhaps as far back as AD 500.[63] During the Gothic age German miners had settled there, bringing with them water-powered crushing machines and other modern equipment. It was an unusual operation, for the copper of the Stora Kopparberg did not run down into the earth in narrow twisting veins, as was typically the case. Instead the mountain was more like one great lump of ore, and from the start it was mined as a large pit open to the sky. The liquation-based revival in central Europe left Stora Kopparberg behind. There were two reasons for this. The first was the paucity of its silver. The second was the political situation within the country. Sweden was at the time virtually a Danish possession, and an economic lethargy hung over the people. Needed investments were not made. Things continued in this way until the early 1520s, when Gustavus Vasa, a Swedish nobleman, won back his country from the Danes. The monarchs of the House of Vasa were to lead Sweden onto the centre of the European stage.

Gustavus was a tyrant and a nationalist and set about strengthening his country. Early on he recognized the value of metals both as exports and a source of weapons for his army. He encouraged the struggling iron-makers in the northern district of Dannemora, stepping-in with funds to help build the larger-scale works typical of the Continent. He took an interest in the technical aspects, at one point questioning the judgement of his ironmaster, Markus Klingenstein, when the German proposed placing two water-powered hammers in a single building: 'you [will need] to build a house large enough for a cathedral'.[64] Responding to his enthusiasm the industry grew in confidence, and soon it was flourishing. Copper, too, benefited from his patronage. He established near the copper mountain a smelting complex equipped with all the latest machinery. The open pit arrangement meant that mining was considerably cheaper than it could be underground.

Opportunity knocked when the Spanish monarch decided to jettison his now hopelessly inflated silver currency and replace it with copper coinage. The weakened and deep German mines could not respond to his demands, and internal political strife was too much of a distraction in Hungary. The Spanish copper was supplied from Stora Kopparberg. It was the cheapest source available, and the only one that could supply such a large quantity in a short time. To help meet the Spanish demands, the mining equipment was modernized with the help of a German master miner, Christopher Klem, who installed better pumps and more efficient hoists for bringing the ore to the surface. With the recoinage complete the expanded production of the great pit had nowhere to go but into markets already being supplied by Bohemia, Saxony and Hungary. Lower costs allowed the Swedes to muscle in. Stora Kopparberg became a formidable force on the European market.

It was Sweden's most glorious age. When Gustavus's grandson, Gustavus Adolphus, came to the throne in 1611 he proved himself to be fully worthy of his forebear. He set out to create a Swedish empire, and by the end of his reign had collected a series of large possessions in Germany, Scandinavia and in the lands bordering Russia. Where did the money come from to finance his empire-building? The answer can be expressed in one word: copper. Indeed, the copper mountain was to provide the wherewithal for 50 years of Swedish international adventurism. At one point it supplied an estimated two-thirds of European consumption. Stora Kopparberg's natural cost advantage helped achieve this dominance, but so too did the Walloon nobleman and merchant Louis de Geer. He it was who sold the Swedish copper on the Rotterdam market, deftly manipulating the price and always staying one step ahead of the game. The competition was weak. The German mines were closed or struggling, and in Hungary a revolution against Habsburg rule had left the smelters and all the surface equipment at Neusohl in smoking ruins. When production there had recovered, Gustavus's capable minister, Axel Oxenstierna, urged his monarch to consider an alliance with the Hungarian nobleman who now controlled the mines. For Sweden and Hungary, wrote Oxenstierna, are the only two sources of any importance left on the Continent.[65] The copper that could not be absorbed in Europe found its way into ships bound for the Indies. There it met competition from another source: Japan.

The discovery of the Ashio mine marked the beginning of a period in which Japan rose from being a self-sufficient copper producer with a small export surplus to a major supplier to the world. The Ashio silver–copper find came just a short time after the introduction of liquation, and when it is considered that the Osaka *Saigerhütten* complex reportedly provided work for 10,000 people, it seems at least possible that the process had an impact as great in Japan as it had in central Europe. With powerful sponsorship from the Tokugawa shoguns, production grew, and exports found their way first to China then further afield to south-east Asia and to India. It was in the Indian market that the Japanese copper came into competition with the Swedish variety. This was a managed competition, for the sellers of both were the officers of the Dutch East India Company, whose sailors had fought and defeated the Portuguese for supremacy in Asia. Japanese copper seems to have had the upper hand in the tussle, and following a calamitous cave-in at Falun it gained a foothold in the European market. Its purity allowed it to retain its hold, even after full recovery at the copper mountain.[66]

From 1600, for a century, Japan and Sweden would dominate the copper supply of the world, just as the Americas, with the help of Japan, dominated its silver and gold. Trade in metals was now global, voluminous[67] and sophisticated. Yet during the period, the total production of the industrial metals changed little, and the global output of silver actually fell.[68] Potosí, indeed, entered a precipitous decline almost as soon as the tailings were exhausted, while in Japan the production of silver had fallen so far by 1668 that its export was banned. In Mexico, only a continual process of discovery made up for the exhaustion of older fields. The period, in stark contrast to the Renaissance era, saw few important technical innovations in the metallurgical industries. It was as if mining and smelting activity had reached a plateau. It would take a series of revolutionary industrial advances to propel it to greater heights.

TWO

The Watt Engine

THE ENGLISH ASCENDANCY

Throughout the Fugger era the mining scene in Britain had remained fairly quiet. This is not to say that it was inactive. The tin of Devon and Cornwall was of global importance, and the production of lead, though small-scale, was healthy. In the first half of the sixteenth century Henry VIII had encouraged and fostered iron-making in the forests of Kent. The trouble was that the kingdom had none of the glamour metals – gold, silver and copper – that were causing so much excitement on the Continental mainland. Henry's daughter, Elizabeth I, sought to redress this lack by creating the kind of state-controlled and sanctioned industry that was enjoying so much success within the Habsburg lands. This she did by establishing two companies to organize the prospecting, mining and smelting of copper and precious metals. They were the Company of the Mineral and Battery Works[1] and the Company of Mines Royal.[2]

The Mines Royal story began in 1563 when Daniel Hochstetter, a representative of the German firm of Haug, Langnauer & Co., visited Elizabeth's court and was able to obtain from her Privy Council the exclusive rights to seek and mine copper, silver, gold and mercury in Wales, Devon and Cornwall and in many of the western counties of England up to the border with Scotland. Hochstetter was the son of the Joachim who had been appointed by Henry VIII to the position of principal surveyor and master of the mines of the realm. With the help of his experienced countrymen Daniel eventually struck paydirt at a place near Keswick in the north-western county of Cumberland and proceeded to develop a mine. He gained permission to establish a drinking house there that served good German wine rather than the bitter beer favoured by the locals. The foreigners were not made welcome by the Keswick townsfolk. Nevertheless, after twelve months good progress was being made in both the digging of ore and in the construction of the smelting facilities.

It is at this point that Elizabeth's plans for a state-controlled industry met a stout challenge. The local landowner, the Earl of Northumberland, indignant at what he saw as trespass, sent in his retainers to seize the nearly 30 tonnes of newly dug ore. The Earl claimed that as the mine workings and the soon-to-be-built smelting house were situated on his land, any produce from them must, perforce, legally be his. The Crown, in its turn, brought a lawsuit against him, claiming damages for the loss of taxation revenue due to the Queen. The case went to trial, and in the event the justices of England decided against the Earl. Their reasoning was that as mines of gold and silver were, by law and long tradition, the property of the Crown, the produce of Keswick must clearly belong to the Crown. Only one detail flawed an otherwise model proceeding: the ores of Keswick were moderately endowed with copper, but contained merely traces of silver and no gold at all! The only – somewhat lame – explanation that has been offered for this odd outcome is that the counsel for the Earl mistakenly believed there was gold at Keswick. The judgement has passed into history not for its primary result but for its secondary ruling. Further down in their summation, the justices asserted that the common law handed down through the generations held that while gold and silver may be the sole preserve of the sovereign, the produce and profits of mines of copper, tin, lead and other like metals 'shall pass to the owner of the land'.[3] It was a marked departure from the Regalian laws prevailing elsewhere in Europe.

Even this secondary ruling would be of only passing interest were it not for another crucial fact. Two hundred and fifty years after the verdict was handed down, the British Isles boasted a mining and smelting industry that far surpassed that of any other country on earth. It was the world's leading producer of iron, copper, tin and lead. Some have suggested that the Keswick verdict, by underpinning a more liberal mining regime than existed elsewhere, was instrumental in this phenomenal success. The argument is a dubious one. For a start, the greatest era of central European mining unfolded within a tight framework of state control. Further, if the ruling did indeed launch British metal on its path to world leadership, we are entitled to expect an upward trend in production throughout the seventeenth century. Instead we see a decidedly mixed picture. The lead industry did in fact undergo great expansion, but in iron and tin the pace of growth can best be described as glacial. Copper production actually fell, thanks to the closure of Keswick. The claim that the Keswick ruling led to more metal can be safely rejected. Unfortunately this does not settle the matter. What, then, *was* the explanation for Britain's metallurgical ascendance?

The answer lies to some extent in the superiority of her ore reserves, for there is no doubt about the richness and accessibility of the English deposits of tin and lead, but this is far from the full explanation. The British Isles held no such natural advantage in iron or copper and, in any case, such an explanation provides no help with understanding the reason for the uneven growth during the seventeenth century and the much faster pace during the eighteenth. To explain the British ascendance we must turn to the industrial revolution. It conferred two benefits on the metallurgical industries. First, the industrial advances themselves – many of which were pioneered in the mines and smelters – lowered their cost of production. Second, the growing industrial base created a buoyant demand for metals. All of which raises two further questions: why did Britain lead the industrial revolution, and how did it hold its advantage for so long? These matters have been much debated and no firm answers have ever been settled. There is, however, consensus on two of the primary factors.

One of them was the British Parliamentary system and the rule of law, protection of property, guarantees of personal freedom and the politics of consent which that institution embodied. This is not only a modern interpretation. This view was held by the *encyclopédistes*, heralds of the French Revolution of 1789. They were much impressed by Britain's Glorious Revolution of 1688, when the incumbent sovereign, James II, was bloodlessly ousted in favour of his Dutch son-in-law, William of Orange. William's accession opened the way for the development of the Parliamentary system as we know it today. It was a Parliament of merchants, and the enterprising businessman was encouraged. Out went the internal tariff barriers. Out, too, went the moribund Elizabethan monopoly companies and the royal right to the gold and silver won from domestic mines. A second factor supporting British industrial pre-eminence was her colonial empire.[4] By 1760, Britain was the main imperial power in North America, the Caribbean and India. The London-based East India Company was as dominant in oceanic trade as Portugal and the Netherlands had once been. The large American and Asian markets were thus preferentially awarded to British producers. As the British Empire grew in strength and reach, so too did her manufacturers. This combination of a favourable political environment and an advantageous position in world markets, it seems, helped the British people to be more successful than anyone else at seizing the technological opportunities that were there to be had. Of these there were plenty beginning to pile up as the eighteenth century dawned. Let us examine the three that were to have the greatest impact on mining and smelting in the next 100 years.

BLASTING THE ROCK

One of the illustrations in *De Re Metallica* shows a miner holding a cloth over his nose while behind him flames leap from a pile of wood stacked up against the rockface. The illustration depicts the technique of fire-setting. This involved lighting a fire in order to heat the rock to a point where it would crack and thus become much easier to chip off the wall. It seems that fire-setting was used only when the rock had become very hard to break. The reasons for its sparing use are clear enough. It filled the narrow galleries with smoke, thus ceasing all activity for perhaps eight or sixteen hours, and it used a great deal of wood while wasting most of the heat. Clearly there existed a great incentive to find a better way of dealing with harder varieties of rock.

The first definitely known use of explosives in mining is dated from 1574 at the lead and silver mine of Schio in the Venetian hinterland. A certain Giovanni Martinengo pioneered the technique, claiming that his new method would 'extract ... greater benefit than the mind of any man has ever imagined'.[5] The path to major technological breakthrough is never smooth, and the Schio exercise was clearly a failure. No doubt there were many who ridiculed the idea. The disapproval is evident in a report written twenty years later by the Inspector-General of the Venetian mines. In the report he reflects disdainfully on the experiment: 'this man did not work his mines ... in the ordinary way of working, but, in an eccentric fashion, making a little hole in the mountain rock with artillery powder, wished to open up and shatter the mountain by force, and thus discover whatever was within that had hitherto remained undiscovered'. Far from being eccentric, it seems surprising that the idea of blasting should have taken so long to germinate. After all, the use of gunpowder for military purposes had been widespread since cannon had become a standard of warfare. Its use underground in military applications was also long-established. Army engineers would sometimes be directed to hollow out a cave beneath a potential battlefield and place in it a dozen or more barrels of gunpowder. They would then be detonated at the most opportune moment. These military 'mines' were very often prepared by experienced miners.

To find the next recorded use of underground blasting in a mining operation it is necessary to cross the Italian Alps. Here we discover that at Thillot in the northern French mountain range of Ardennes an order was placed in 1617 for just over nine francs worth of 'powder for exploding in the mountain and moving rock'.[6] Thillot provides the first known example of the regular and large-scale use of explosives in a mining operation. The reason why gunpowder was employed at Thillot can be surmised from the surviving records. These show that work there had ceased entirely, probably due to the hardness of the rock. It was a common problem. Very hard rock is described by Agricola as a major impediment to hewing. He tells of some silver-rich galleries in Saxony in which the rock became so hard that progress of a few inches a week, or even a month, was all that was possible. The introduction of gunpowder was driven by the desire to restart the mine.

The next recorded use is just south of Thillot. Then, in 1627, we find it being applied for the first time at the thriving silver mines of Schemnitz in Slovakia. The pioneer at Schemnitz was Caspar 'Blaster' Weindl, a native of the Tyrol. Weindl had travelled from Florence to Schemnitz in order to put into practice his ideas for mine blasting that, according to him, he had conceived while he was serving as an officer on campaign in Florence. He makes no mention of Thillot, and there is every possibility he arrived at the insight independently. The galleries in which he first tested his technique had been abandoned for years, the rock there being notoriously hard to cut through. The management's main concern was that the blast would disturb the timber supports throughout the underground passages. In the event their concerns came to nothing. The official report on the blasting trial records that it was a complete success.[7] The result of the innovation was a virtual doubling of silver output. Gunpowder had made its mark in one of Europe's premier mining districts.

From Schemnitz blasting spread across Europe, although its acceptance was neither rapid nor universal. The main obstacle in the way of its adoption seems to have been the danger involved in using it. A report still survives of an incident in the first year of gunpowder use at the silver mine of Gastein in the Tyrol. A worker experienced in explosives had visited from a neighbouring shaft to help with the introduction of the new technique. He hammered two holes into the rock face and then packed them with explosive. He then proceeded to plug the holes using the iron wedges that were then in vogue. Completing the first hole, he started on the second. Probably a spark flew. With no warning the charge exploded in the face of the unfortunate man, throwing him against the opposite wall of the gallery and severely burning him and injuring two of his companions. Incidents like these did much to deter miners from even attempting to work with the unpredictable powder.

Nevertheless, its triumph was inevitable, and by the end of the century it was in common use throughout Europe. Given its potential to increase the volumes of rock that could be moved in a day, it is curious to find that the introduction of blasting made little immediate impact on the overall output of the mines of Europe. So why was it so widely adopted? The answer seems to lie in the effect it had on the costs of mining. The most obvious benefit in this respect was the reduction in the number of people needed in the underground workings. Some have suggested that a second cost benefit came about because it allowed substantial reductions in the consumption of wood.[8]

SMELTING WITH COAL

The forests of Europe had been shrinking since medieval times. The spread of agriculture was the major cause. Also, as the population had grown, so too had the demand for firewood and for timber for housing. If the surviving historical records can be relied on, England may have been the first country to experience real pressure on its forests. By the time of Elizabeth's reign a scarcity of wood was beginning to make itself felt. Her Parliament passed legislation restricting the felling of trees within fourteen miles of navigable rivers.[9] These forests were a vital source of wood for shipbuilding. Next we hear of a Coventry city ordnance prescribing the stocks for anyone caught taking wood from the forests. A second offence meant expulsion from the city. In this sort of legislation can be seen the first blows struck in the 150-year struggle for a sustainable and low-cost alternative to wood fuel.

Of alternatives there was really only one. It was the mineral that Marco Polo had described as the 'shiny black rock'. Coal had been used in Europe for centuries, and in China for far longer. At an early date 'sea coales' were being shipped from Newcastle upon Tyne to London for use in the hearths of poorer townsfolk. It was never a very popular fuel, and was shunned if the alternatives of wood and peat were available. The problem with coal was its acrid smoke. Unlike wood, it contained plenty of sulphur and other impurities that were released when it was burned. The choking fumes that resulted were a strong deterrent against its use in domestic heating and cooking. Its range of industrial uses was also limited. In some industries coal and wood were interchangeable. The manufacture of gunpowder, for example, required saltpetre that in turn was prepared by boiling a vat containing animal excrement, bird droppings and various other ingredients. The switch from wood to coal in this process presented few technical problems and could hardly worsen the odour. Salt manufacture was similarly indifferent.

In other industries coal could not be used. In the drying of hops for brewing, for example, the sulphurous smoke left the beer with a bitter and distinctly unappealing taste. In the case of metal smelting, the sulphur that remained trapped in the metal left it even more brittle and severely restricted its range of application. Glass that was produced with coal emerged cloudy and uneven. Thus wood remained in high demand, and its price rose to the point where its cost became a competitive disadvantage in many industries. Nowhere was this more acute than in English glass. The Flemish, Italian and French manufacturers were low-cost competitors, and the high price of English wood threatened to extinguish the fledgling industry just as it had gained a foothold in its home market. The problem was solved when Thomas Percival and Thomas Hefflyn devised an enclosed crucible wherein the potash and sand, the raw materials in the glassmaking process, were protected from the smoke.[10] Their technique saved English glass. It no doubt also got others thinking.

The quantity of wood required for smelting was huge, and the industry's gluttonous consumption aroused resentment as well as presenting the smelter owners with a mountainous bill. An added incentive to use coal existed for the iron-maker and the lead-smelter in that very often their deposits lay close to rich coal seams. At last, in the final quarter of the century, a series of innovations resulted in the ability to use coal in the smelting of copper, lead and tin. Chief among these innovations was the reverberatory furnace (so-called because its domed roof 'reverberated' the heat from the burning coal onto the ore without allowing the smoke of the coal to come in contact), which was mastered in the south-west of England. Its first successes were achieved on lead ores by the Bristol-based Clement Clerke. The patent application, which dates from 1678, claims the invention of a process 'to melt and refine lead oare, in close or reverberatory furnaces with pitt coales, turfe, peate or other mixt fuel and not with woode, woode coale, or charcoale ...'.[11] Lord

Grandison, in submitting the patent, clearly stated that the benefit of the coal lay in the fact that it avoided the 'dearness and scarcity of wood'. However, Clerke and his two assistants, John Coster and Gabriel Wayne, were unable to bring Lord Grandison's lead works into profit, and after a few years parted company with him.

Having learned a good deal about how to use coal in the reverberatory furnace, and with Grandison's patent preventing them from working any more with lead, the trio turned their attention to copper. While Clerke and Wayne travelled to London to set up an experimental smelter, Coster returned to his native Forest of Dean in Gloucestershire, where he knew of some small copper deposits. He set up his reverberatory furnace in the small village of Redbrook.[12] It prospered, and was soon joined by another firm. Coal-based copper smelting had become a reality. Given the relative closeness of Gloucestershire to Cornwall, we should not be surprised to see the reverberatory furnace also applied to tin at around the same time. The pioneer in this case was Robert Lydall. The letter granting him a patent for 'A New Way of Smelting and Melting Black Tinn' cited the 'great benefitt to the Publick'[13] of hugely reducing the wood used in the smelting process.

Even with these three triumphs the problem of coal-based smelting was only half solved. The biggest prize of them all – coal-based ironmaking – still awaited its champion. It would find it in the person of Abraham Darby, an ironmaker in Coalbrookdale, a village on the River Severn, 25 miles north-west of Birmingham. In Birmingham itself could be found one of the success stories of a newly confident England. Its success was built on metalworking, an activity for which its location was well-suited. For by this time the iron industry of England had largely abandoned its birthplace in Sussex and migrated west in search of that happy conjunction of plentiful iron ore, cheap labour and wood, swift running water and coal for the forging. It had settled in the Forest of Dean and in the area to the west of Birmingham. Here it found the natural resources of all kinds necessary to nourish a growing industry. In the decades after the accession of Charles II in 1660, Birmingham rose to a position of prominence. In this it was helped by an Act passed into law in

1665. The Act banned Nonconformist preachers from living in any town that sent its representatives to Parliament. The result was an exodus of Dissenters from towns throughout the length and breadth of England. They were welcomed in voteless Birmingham, and soon the town had acquired a reputation as a bustling community whose citizens had an eye for the main chance. Business was buoyant. When Charles II ascended the throne in 1660, following on from the austere regime of Oliver Cromwell, Britain – and England in particular – enjoyed a resurgence of luxury, and a market for shiny brass ornaments sprang up alongside the existing markets for brass buttons and all manner of iron goods. Birmingham supplied them all.

It was in this environment of metal trades and entrepreneurship that Abraham Darby began his working life. His father was a locksmith in a small town outside Birmingham.[14] He was also a member of the Society of Friends, a Dissenting religious group commonly known as the Quakers. A large community of Quakers lived in Birmingham, and when the time came young Abraham travelled there to be apprenticed to Jonathon Freeth, a maker of malt-mills and an active member of the Society. Having completed his apprenticeship, Darby married before moving south to Bristol, where he set up a business as a maker of malt-mills. He did not stick at this for long. Bristol was a copper town, and it was this expanding and vigorous industry that caught his imagination. With four partners, Darby established the Bristol Brass Wire Company. They mastered the art of brass making, and were soon making pots and pans and all manner of other goods. But it was not enough for Darby. He turned to experimenting with the casting of iron pots and pans. Success was not long in coming. In 1707 he was granted a patent for a new way of casting iron. This sort of independence appears not to have endeared him to his partners. In any case he seems to have been a person who needed to follow his own road.

His next move was to strike out on his own in order to exploit his invention. This he did by forming the Bristol Iron Company. In a short time he had outgrown this as well. The furnace available to him was too small for the volume of business and the size of the articles he wished to pursue. A blast furnace was needed. There happened to be a second-hand one available in Coalbrookdale. It was a coalmining district with good transport links to Birmingham and Bristol. It was also near his home town. Darby moved there, keeping open his Bristol works but concentrating on getting his much larger new works up and running. He would remain hard at work in Coalbrookdale until his death there at the age of 39.

Only much later was the discovery made that Darby was the author of one of the greatest of all the many marvels of the industrial revolution. It was during his early years at Coalbrookdale that he mastered the smelting of iron using coal.[15] The key was to cook off all the impurities in the coal and create what Darby called 'coakes', and what these days we call coking coal. Some believe he derived the idea from the brewers, who by that time were regularly using these specially prepared 'coakes' in the drying of hops. Darby took out no patents and kept his discovery to himself. So little fanfare accompanied the breakthrough that 50 years after his momentous success, the smelting of iron with coking coal was being used by only a handful of other iron-makers. When it finally did become widespread, it was believed for many years that Darby's son, Abraham Darby II, was the inventor of the process. The modern era of coal-based iron-making had begun with a whimper. The bang was to come three generations later.

PUMPING OUT THE WATER

The awakening of a new spirit of enquiry into natural phenomena had been one outcome of the Renaissance. No longer were the handed-down theories of the ancient Greek philosophers deemed sufficient as an explanation of natural things. Here and there individuals began conducting their own experiments, and scientific enquiry began to gather pace. Around 1660 Robert Boyle wrote the pioneering work in which he described and demonstrated the existence and power of atmospheric pressure. Then it was the turn of the Frenchman Denis Papin and his pressure cooker. This was a closed vessel filled with water and equipped with a tightly fitting lid and a specially designed pressure release valve. By boiling the water Papin demonstrated the powerful build-up of pressure inside the vessel. He conceived the idea of using the head of steam thus created to drive a piston. Papin's invention and his speculations on the use of steam power somehow reached the attention of the Englishman Thomas Savery. Savery had worked much of his life as a military engineer, spending his time indulging his flair for invention. He was attracted to the problems of drainage in the coal mines. Noting that the usual methods of water removal from mine shafts – the driving of adits and the use of the endless bucket and chain – were often ineffective, in 1697 (or perhaps in the following year) he built his first steam-powered vacuum pump. He dubbed it the Miner's Friend, and promoted its advantages to mines and collieries. It was installed in only a few places, however, and was found to be inefficient, unreliable and dangerous.

The Savery engine failed because it was poorly designed. The crux of the problem was that it used steam to do too much. The steam was asked to both push the water up a pipe and then, when it condensed in the act of doing this, create a vacuum that pulled still more water up the pipe. It was extremely inefficient of heat, and the explosive reaction that occurred when the steam came into direct contact with the water was both dangerous and damaging. On top of that it did not succeed in raising water very far. It would take an ironworker from Devonshire to do the job properly.

At his workshop in the Devonshire coastal town of Dartmouth, Thomas Newcomen would have heard countless tales of flooded mines in the nearby tin districts. No doubt he visited several of them in the course of his normal work. It is unclear how he first came into contact with the Savery engine but, having done so, his observations on its operation led him to two essential insights. He realized that using steam to physically push water upwards was the wrong idea. He also grasped the fact that the chamber in which the steam was condensed should not be located in the same pipe as the one through which the water was being drawn. In applying these insights, the new machine he created was a vacuum pump, the working principle of which was that the condensation of the steam created a vacuum that in turn sucked the water up the pipe. It was instantly recognizable by its huge crossbeam, which continually see-sawed between the condensation chamber and the pipe up which the water was being drawn.[16] It was also recognizable by the fact that it worked. The first known installation of a full-size Newcomen engine, into a Staffordshire coal mine, dates from 1712.[17] One of the next customers was the Cornish tin mine of Huel Vor. Soon the engines were in use throughout England. Newcomen pump houses with their rocking crossbeams and belching smokestacks could be spied in many places on the rolling hills of Devon and Cornwall.

The final step on the road to the steam engine would be taken by a nervous and irritable Scotsman working at Glasgow University. The 30-year-old James Watt was employed there as a surveyor and an instrument repairman. Having on one occasion been given a model of a Newcomen engine to repair, he had started to think about the ways in which its well-known inefficiency for fuel could be improved. The answer, he realized, was all a matter of heat conservation. The Newcomen engine was built around a boiler chamber, within which water was heated into steam and then cooled down before being heated up again.[18] Watt saw that the job of repeatedly heating up the boiler chamber was the reason for the great fuel consumption in the Newcomen engine. His second insight was to recognize that the steam could be used to provide actual motive power, instead of just being used to provide a vacuum when it was condensed. He proceeded to design a revolutionary pumping engine in which the steam was used to push the piston. When the piston reached its limit, the steam was allowed to escape from the boiler chamber into a separate condensing chamber.

Watt was a skilful mechanic and, as such, popular among the scientists and businessmen who constituted the University's professional community. One of his admirers was a certain John Roebuck, who in turn was friendly with a Birmingham industrialist named Matthew Boulton. Roebuck wrote to his friend, and in the letter mentioned the talented young man who had conceived an improved Newcomen engine. It was the start of a chain of events that changed the world. For Matthew Boulton, owner of the SoHo Manufactory, was a remarkably gifted businessman and not one to let opportunity slip by.[19] He invited the awkward engineer to Birmingham. It was a meeting of minds.

Several years passed after that meeting, during which Watt made several half-hearted attempts to get his pump going at Roebuck's colliery at Bo'ness in Scotland. When at last the colliery closed its gates, Birmingham beckoned, and in 1774 Watt moved there. Boulton lost no time in pushing his protégé to perfect his brainchild. In the meantime he set about establishing a pumping business.

He saw at once that Watt's fourteen-year patent had already run six years, and that with only eight years remaining it was unlikely that any substantial investment in development would be recouped. He petitioned for, and was granted, an extension of the patent for another 25 years, an act that gave the pair propriety over steam-powered pumping until 1800. That done, they teamed up as partners in the firm of Boulton & Watt. While Watt concentrated on correcting the defects of his engine, Boulton began planning how best to bring it to the attention of the industrialists and miners of England. He was an unusually skilled marketeer, with a flair for attracting publicity, and he determined to launch the Watt engine with a flourish. Accordingly, when he received an order from the Bloomfield colliery for a pumping engine, the scene was set for the great unveiling.

Watt was nervous. He had assembled his machine using the best of components – brass fittings from the SoHo Manufactory and a specially machined piston casing purchased from the famous ironmaster John Wilkinson. Still, what if it failed in so public a test? This new engine was nearly three times larger than the one that had performed so indifferently at Bo'ness. In the event, the firing-up of the engine at Bloomfield was a triumph.[20] A second, built to drive the bellows at Wilkinson's ironworks, was likewise a success. Orders began to come in from municipal waterworks, distilleries and coal mines. But in all these applications the existing Newcomen – and even Savery – engines could adequately perform the necessary pumping task. Further, these older engines were cheaper to construct and, in the areas where coal was abundant, not very much dearer to run. In the collieries, indeed, the coal was free. What Boulton and Watt wanted was an application that needed more power than a Newcomen engine could provide and which was concentrated in an area that was lacking in coal.

Boulton knew the only place that would fit the bill.

Cornwall is not a mountainous region, and it is wet. The rainy season continues most of the year and the countryside is checkered with moors and criss-crossed by large streams. Its mines had been pushing deep downwards into the rock for many decades. Drainage tunnels being a practical impossibility beyond a certain depth, the descent to deep levels had been made possible only by the Newcomen engines. For two generations these pumps had done their job, a job continually improved by experience and engineering innovation. The engines were now perhaps as efficient as they ever could be, but still they were not keeping pace with the water. The shafts were getting deeper. On top of this the price of copper was starting to fall. The miners needed more power for less coal. So it was that when Boulton set about wooing these sceptical and conservative businessmen, success was not long in coming. An order came in from the Ting Tang copper mine in the Redruth district. Close on its heels came a second, this one from Wheal Busy near Camborne.

The Wheal Busy components were the first to be readied for assembly, and it fell to Watt to make the journey south to oversee the installation and commissioning of the engine. He and his wife found lodgings in Chacewater, north-east of Redruth. The hardbitten and closeknit community of Cornish miners did not welcome the newcomer, and the sensitive Watt was soon complaining: 'Certainly, they have the most ungracious manners of any people I have been amongst.' His wife chimed in: 'The spot we are at is the most disagreeable in the whole country.'[21] On both counts they were probably justified in their sentiments. The copper mines of the time had become concentrated in the triangle formed by Chacewater, Redruth and Camborne. The rolling hills were pocked with hundreds of shafts, many topped by pumphouses from which coal smoke wafted all the day long. As for the ungracious manners, clearly little had changed since the days of Sir Walter Raleigh, when the miners were said to comprise 'ten or twelve thousand of the roughest men in England'.[22]

When the crowd assembled to watch the boilers of the Wheal Busy engine being fired-up, there were doubtless many among them who were hoping for failure. They were disappointed. The Wheal Busy engine roared into life, and within hours it was apparent to all that here was a pumping engine that was well advanced on Newcomen's venerable machine. The potential for savings in coal became the talk of the area. Orders began to mount up at the SoHo works, and Boulton struggled for workers who could do the careful work of precision machining the valves and other parts that regulated the pumping action. He also searched for suitable engineers who could be entrusted with the task of overseeing the on-site machine assembly and the commissioning of the pump. It was a busy time. Watt spent many months of the year in Chacewater, grumbling and complaining about the inadequacy of the local engineers – they were all, he wrote, drunkards and incompetents – and about the filthy habits and deviousness of the Cornishmen. All the while he strove to improve the efficiency of his pump. One of the challenges was to seal better the small gap between piston and casing to prevent steam from leaking out. Eventually he settled on a thick greasy mixture of tallow and string. It worked, but it had one drawback – the Cornish enginemen ate it, or so Watt claimed. But the greatest trial for Watt, and even more so for his partner, was the securing of payment from the mine owners who bought the engine. Many of them were close to bankruptcy, and certainly they were all close to their money. One of the main reasons for the poor financial performance of the mines was that in the cut-throat world of business, they were the ones getting their throats cut.

THE ASSOCIATED SMELTERS

In the hundred years before Watt's visit to Cornwall, the tin mines of that county had greatly benefited from the new technologies of gunpowder and the Newcomen engine. These were the weapons with which the miners could defeat the rock and water that stood between them and the rich tin that lay at depth. As the shafts had pushed further underground it was found that in the lower depths lay good reserves of copper, and over time this metal grew to be more important than tin. While the ancient Stannary laws[23] made it mandatory to smelt the tin in Cornwall or Devon, no such restriction was placed on copper, and from the start it was smelted in coal-rich South Wales. John Coster's coal-fired smelting furnace had proved a great success. The first generation of the furnaces had been built around Bristol, and a merchant navy had become a feature of the Cornish coast, transporting the ore from the inlets of Cornwall to Bristol. On their return journey they carried coal for fuelling the Newcomen engines and tin smelters. As the industry grew, the centre of gravity of copper smelting shifted from Bristol to Swansea on the coast of South Wales. Among the attractions of this new site was its closeness to Cornwall, a good harbour and, most importantly of all, proximity to the abundant and cheap coal supplies of the Welsh valleys.

The departure of their ore to Wales was a fact that did not escape the attention of the Cornish miners, and they came to regard the smelter men with suspicion and resentment. Typical of the sentiment was the following lament: the smelters 'take care to keep as much in the dark as they can by shipping off all the ore to be smelted in their houses near Bristol, in Wales, etc., under a pretence of saving in fuel; but in reality to increase the profit ...'.[24] It was a reasonable suspicion. From the beginning the Swansea smelters had found it convenient to band together in their purchase of ores, and over time a highly controlled system of ore purchase developed. Under this system the smelters would each submit closed bids, or tickets, for a parcel of ore produced from a particular mine. The highest bidder would get the ore. In this bidding process the smelters routinely colluded, knowing that the miners were obliged to accept one of the bids, no matter how unsatisfactory the price may be. The success of 'ticketing' encouraged the smelters to organize themselves into a more formal cartel arrangement. It came to be known as the Associated Smelters.

By mid-century it was a group of nine family-owned concerns.[25] Most of them were situated either along the banks of Swansea's River Tawe or nearby; a couple of others had stayed in Bristol nearer to the brass works of that city. Things had progressed a long way in the 60 years since John Coster's first experiments. The smelters were now large and technically sophisticated. All of them used a variant of the Welsh process, a complex series of nine interrelated furnace-based operations. Not much is known of the stages in the development of the Welsh process, in part because the experiments and trials that attended its evolution were undertaken in great secrecy. The earliest known detailed description is in the form of a flowsheet attached to an 1811 agreement between John Vivian, a leading member of the Association, and William Howell, a newly appointed smelter manager. The agreement constrained Howell never to pass on the technique except to his children.[26]

The Welsh process provided the Associated Smelters with the ability to produce cheap copper of very high quality. The core of its advantage lay in the preparatory steps prior to the actual smelting process and in the careful blending of different types of ore and partially processed metal. Its use of inexpensive coal offset the relatively high cost of the copper ore. In a Swedish or German works, a batch of metal might have to be re-smelted ten times or more before it was of a purity to allow the refining stage to begin. The Welsh process reduced the number of these repetitions, saving substantial cost in doing so.[27] Needless to say, the science of this involved procedure was a complete mystery to the furnace operators, and the Swansea smelters came to rely on experienced men. They were much sought after and jealously guarded. The complexity and intricacy of the process prevented others from copying it. Secure in their technological advantage, the members of the Association concentrated on obtaining cheap ore and cheap coal, and on selling their copper at healthy prices. A tariff wall protected them from any threat of foreign competition.

Apart from supplying the multiplying wants of the domestic market, the Swansea smelters also held a privileged position in the overseas trade. Great Britain now ruled the waves, and its merchant fleet dominated the trade routes both east and west. To the islands of the Caribbean were sent copper cauldrons for making sugar and rum. British sea captains traded copper for slaves along the coast of West Africa. The London-based East India Company sold copper bar to India for ornaments as well as for stamping into coins. The India trade was a prize the

Association valued above all else, and they were prepared to offer generous concessions to retain the favour of the East India Company. Prices charged to the Company were up to twenty per cent lower than those charged in the home market.[28] Not that foreign competition was particularly fierce. Stora Kopparberg was now merely a minor producer for the Baltic region, and Japan was a spent force on world markets. Only the thriving Russian industry (discussed in the following chapter), northern Africa and a rump of German mines provided any real alternative for the customer.

Adding to all this was the emergence of a new and lucrative market. British shipowners were increasingly turning to the use of copper sheathing, or 'copper bottoms', for the protection of the wooden hulls of their vessels.[29] Wooden ships had always suffered the problem of encrustation of their hulls by seaweed. Left unchecked the seaweed would slow the ship and hasten the rotting of its wood. Ocean-going vessels were frequently put into dry dock for cleaning and for the replacement of their hulls. As British ships made more frequent and longer stays in the warmer waters of the Indies, these problems were exacerbated. The Royal Navy began seriously to seek a solution. Copper sheathing of the hulls was tried. It was a controversial experiment and one that seems to have failed. The copper wore out and, worse, the iron bolts and fastenings used in the hulls tended to corrode when in contact with the copper. This had the effect of putting ships in quite literal danger of falling apart. Despite these initial setbacks, the use of sheathing gradually became more effective and more widespread. Then came the American War of Independence. In 1778 George III authorized copper bottoms for the entire Navy.

THE COPPER KING

Amid all this growth the mines of Cornwall fared less well. The ticketing system ensured that the revenues flowing their way, especially to the poorer mines, were barely enough to cover running costs. The richer mines were more comfortably placed but rarely accumulated a cash surplus, given their custom of distributing all profits in annual dividends. So little provision was made for the future health of the business that the mines developed in a curiously haphazard way. The managers did the best they could. Their dependence on the expensive Newcomen pumps, however, meant that they could not cease the sale of their ore in order to better bargain with the smelters. Ceasing their sales would mean that they could not buy coal to drive the pumps. Crushed at the bottom of this ruthless commercial pecking order were the long-suffering working men.

As Matthew Boulton familiarized himself with the conditions within the industry, it became clear to him that the mine owners could never afford to pay outright for the new pumping engines they so badly needed. He devised, instead, a system of payment by which the partnership would receive from each mine an annual payment based on the saving in coal resulting from the use of the Watt engine. The only initial payment was for the purchase of the engine components. This 'buy now, pay later' scheme was the incentive needed to convince the slower of the mine owners to move. The old Newcomen engines were replaced. In 1784, it is said, only one remained in operation in all Cornwall.[30] Boulton was typical of the breed of innovative and independently minded businessmen who provided the hard commercial edge of the industrial revolution. Not long after he and Watt began installing their pumping engines, Boulton would become a central player in a struggle for supremacy in British copper.

It began on Anglesey, an island separated from the coast of North Wales by the narrow Menai Strait. At the northern tip of the island, overlooking the Irish Sea, lies the small port of Amlwch, and just inland is Parys Mountain. It was there that in 1779 an energetic lawyer named Thomas Williams gained control of a large, somewhat low-grade, copper deposit that lay easily accessible on the side of the mountain.[31] He began mining the copper ores and selling them to Swansea. Modest ambition and ability were never part of Williams's make-up. Still, it is likely that his ambition at first amounted only to a desire to be treated as an equal by that powerful group of industrialists, the Associated Smelters. Everyone knew that the Cornish were being shortchanged. Everyone also knew that the smelters continued to get away with paying unfairly low prices because of the ingrained inability of the mine owners to act in concert against even so obvious a wrong. So it was probably with confidence that Williams approached the smelters to request a higher price for his ores. He was brusquely rebuffed.

It was a mistake the smelters would live to regret. Fired into action, Williams expanded his mining operations and then used the money from the ore sales to build a smelter near Liverpool and a fabrication works in Holyhead on Anglesey. Next he boldly moved into enemy territory by purchasing the Upper Bank smelter in Swansea. He then made a grab for the most precious possession of the Swansea cartel – the East India contract. This was too much. The Association, shaken out of its complacency, responded with grim determination. They cut prices savagely to win the contract, passing all the reduction straight on to the Cornish miners. Undeterred by this defeat the Anglesey lawyer next set his sights on the contract for sheathing sales to the Navy. This he won by helping solve the longstanding corrosion problem with the bolts that fastened the copper sheets to the ship. It was a major coup, and now Williams was a force to be reckoned with. But he did not have the strength to truly test the cartel. They were well-organized, well-connected and wealthy. To achieve his ambitions of market dominance he needed an ally, and to this end he fixed on Cornwall, the source of over two-thirds of England's copper. The Cornish mine owners, caught up in the battle between Williams and the Associated Smelters, were facing ruin. If Williams was to use them against his enemies he had first to save them. So began the enterprise called the Cornish Metal Company.

The idea was simple enough. Instead of merely selling their ores to the Associated Smelters, the Cornish miners would sell only on condition that the refined copper was sold back to them at an agreed price.[32] The scheme was fleshed out by Williams and quickly gained great favour in Cornwall. Not everyone was caught up in the enthusiasm. Richard Phillips, one of the wiser heads of the industry, dismissed the scheme as pie-in-the-sky. He maintained that the mine owners had neither the money nor the skill to succeed in such an undertaking and, in any case, they could never maintain the unity and fixity

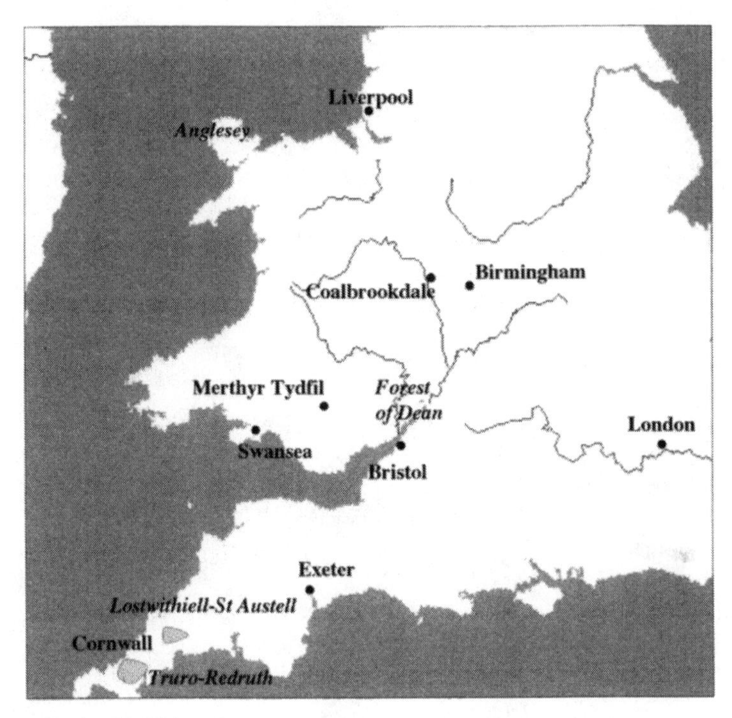

3 England & Wales, 1800

of purpose that the scheme required. He was ignored, and that summer in the Cornish town of Truro a great gathering of mining men, smeltermen and copper traders assembled from all parts of England. Meetings were held. The final negotiations were dominated by Williams and the ironmaster John Wilkinson. The pair of them, according to Boulton, 'drove the Cornubians and Bristol men before them like sheep'.[33] Williams was unimpressed by his soon-to-be allies and 'told them in a full meeting that ... he had no idea of their being half so ignorant as he found them'. But at last the deal went through. Under its terms, Cornwall was effectively turning the tables on Swansea and forming its own cartel. That, at least, was the theory. The deal with Williams was essentially one of dividing up the market. Cornish copper was to have South Wales to itself, while Anglesey kept the industrial region of Liverpool as its own. The two biggest markets, London and Birmingham, were to be shared.

The Cornish Metal Company was soon in deep trouble. No-one on the Cornish side had any aptitude or even, it seemed, interest in doing the hard work of seeking out customers. Previously this had been the work of the Associated Smelters. Williams, by contrast, was everywhere with new products and new customers. While he prospered, Cornwall was soon back in its familiar state of desperation. Unrestrained overproduction led to the predictable result. Mines were forced to close as ore prices fell. The workers were laid off, and this led to rioting, with blame being directed at the Metal Company, and threats made to burn down the house of its chairman, John Vivian. While this disturbance subsided, the pressures mounting on the Metal Company did not. At last, in their despair, the mine-owners turned to the only man they knew who could lead them to safety. So it came about that Thomas Williams found himself at the head of an organization that controlled the sale and smelting of virtually all of England's and Wales's copper. As one member of the Anglesey camp put it, 'every ounce of copper produced by Cornwall is to be sold to Mr Williams for five years and no other man upon earth is to sell an atom of it ... Anglesea triumphs, the command of the trade will be totally there ...'.[34] Williams controlled the lion's share of the global copper market, hence the soubriquet bestowed on him: 'the Copper King'. It was quite an achievement for a man who only ten years before was a little-known lawyer representing one of the parties squabbling over the ownership of a windswept hill on Anglesey.

Williams did not long savour his triumph. The Birmingham industrialists fought back against his high prices, and in time the Associated Smelters regained the initiative. Worse still for Williams was the fact that his copper mountain began to give out. In their struggle to defeat Williams the Swansea cartel had even allowed a certain precarious prosperity to return to Cornwall. Fickle to the last, the Cornish now looked on the Associated Smelters as their saviours against the 'arch-monopolist'. Their dislike fell even more heavily on Boulton and Watt. They were not a popular partnership in the mining districts. The miners accused them of demanding extortionate dues on their engines and of being in league with Williams. When the Copper King took the sensible step of shutting some of the Cornish mines, rioters attacked not the property of Williams but the Chacewater offices of Boulton & Watt. The firm's staff were forced to retreat, with their account books, to the better-policed Truro. Boulton and Watt were not personally threatened. They were already back in Birmingham.

DARBY'S INVENTION FINALLY TAKES OFF

As the frenetic activity in Cornwall settled to a routine, Boulton had urged Watt to consider anew his abandoned plans for the development of a steam engine that could provide a rotating motion, instead of the vertical action that worked the Cornish pumps. Thus challenged, Watt returned to Birmingham and once there embarked on his most productive period. He perfected the rotative steam engine, employing designs of brilliant elegance and simplicity. Forever the pessimist, Watt could not see any real practical application for his new device. The domestic manufacturing industry, he was sure, was in decline and would soon be swamped by imports. Boulton was the better judge of such matters. He knew the optimism and energy of his fellow industrialists.

One of the landmark inventions of the industrial revolution was the Arkwright spinning jenny. This simple device allowed the mechanization of the clothmaking process, and in so doing completely changed the structure and economics of the textiles industry. For a while the new spinning factories were held back by a lack of power to drive their looms more rapidly. It was a heaven-sent opportunity for Boulton. He set about demonstrating to the cotton loom owners the benefits of his rotative steam engine. Other industrialists took note and the Watt rotating engine was put to use driving the looms and all manner of other machines of industrial England. The steam-powered spinning jenny was the foundation stone on which was built the formidable British industrial complex that provoked later writers to describe Britain as 'the workshop of the world'. The spinning jenny and the steam engine were both constructed of iron and so, increasingly, were the factories in which they were used. A demand for iron swept the country, and alongside the textiles factories grew up the iron works.[35]

There is probably no greater symbol of England's industrial revolution than the iron works. It has inspired paintings such as Philip James de Loutherbourg's *Coalbrookdale by Night* (1801),[36] which depicts the flames from a dozen furnaces lighting up the mills and the night sky as great clouds of dense smoke waft overhead. William Blake responded too, with his unforgettable characterization of the factories of England as 'dark satanic mills'.[37] It comes as something of a surprise, then, to learn that prior to *c.* 1780 Britain imported more iron than was made within her shores. First Sweden, then Russia, was the main source of iron for the world's leading economy.[38] To understand why this was so it is necessary to survey briefly the development of the industry following the pioneering use of coking coal by Abraham Darby.

Coke-based iron-making had remained confined to Darby's works at Coalbrookdale and to a few other mills in the surrounding area. The reason was that the metal so manufactured had limited use. It was still too brittle and good only for use in heavy castings. The rest of the industry relied on traditional charcoal-based smelting. The trouble was that these furnaces continued to be small and labour-intensive, managing to flourish only wherever there was a payable iron ore deposit and access to plenty of cheap labour and wood. In this unsophisticated state the industry was as vulnerable as it always had been to competition from lands that had good ore deposits coupled with cheaper labour and cheaper wood. Hence the success of Sweden and Russia. Into the bargain their iron was exceptionally pure and, even when subject to import duty, was preferred by many manufacturers. During the period between 1775 and 1790 all this changed. First, Watt's steam engine appeared on the scene and imparted more power to the furnace bellows. This allowed larger blast furnaces to be built with the attendant reductions in cost.

A new breed of ironmaster emerged. These men built large coke-based works, ran them cheaply and then competed aggressively on price. Their success began to turn back the tide of imports. They could now manufacture iron at lower cost, but the job was only half done. The iron remained brittle. Some tried finding new uses that would tolerate the brittleness. The famous Iron Bridge over the River Severn at Coalbrookdale (1779–80) was one such attempt. But it was by now clear that only some technical innovation in iron-making would provide the long-term answer, and the new mills became host to trials of all kinds. At last an ironmaster named Henry Cort discovered the secret – it came to be known as the puddling and rolling process.[39] In traditional iron-making, the molten metal from the blast furnace would be drained into sand furrows. The main furrow acted as a distribution canal for the iron to flow into smaller furrows lying at right-angles. The arrangement had the appearance of a sow suckling a family of piglets, and so the cooled blast furnace product became known as pig iron. The pig iron would be taken to the 'finery', where it was alternately heated (in order to soften it) and beaten using steam-driven hammers. The hammering helped remove some of the impurities and also served to form the metal into its required shape.

Cort's technique changed all this. To understand his process it is first necessary to picture the molten pig iron after it has been drained from the blast furnace. Instead of being directed into a sandy furrow, it was collected in a thick iron tank. While it was sitting there slowly cooling, it had the appearance of a 'puddle' of molten metal. This puddle was stirred with a long green log, with the result that it 'spits out in blue sparks the sulphur which is mixed in with it'.[40] The scene described is so similar to accounts of the final stage in the Welsh process that it seems reasonable to conclude that Cort borrowed his puddling idea from the smelters that lined the River Tawe. He then took this semi-solid congealing iron and, instead of hammering it, pressed it between grooved steam-driven rollers. The remaining impurities were squeezed out and the iron neatly shaped into bars ready for sale. Observers pronounced the metal so produced to be better than the Swedish variety. The ancient art of iron-making was freed once more from its chains of charcoal.

When Cort patented his process it was stolen by his rivals. On one occasion he was visited at his Fontley works in Hampshire by two industrialists from South Wales. This light-fingered pair returned home not only with the plans of the new plant in their travelling cases but with some of the trained workmen in tow. In an effort to justify their actions these men and others like them sought to belittle Cort's triumph. They claimed he had progressed merely a small step from a process that had been in use at Coalbrookdale twenty years before. Their cause was helped by a certain Peter Onions, who could prove that he had, for several months prior to Cort's patent claim, been operating a very similar puddling process. Perhaps Onions, too, had visited Swansea. Further help in the fight against Cort arrived when a dusty old patent was rummaged up from some forgotten archive. It showed that one John Payne had 50 years before devised a rolling process that, though never actually operated, included the same principles as Cort's. In the end the ironmasters were able to stop throwing mud when, in an episode unrelated to their efforts, Cort's career came to an abrupt end. His demise came about when his backer – his uncle, Admiral Jellicoe – was found to have raised the finance for his nephew's large new works by dipping into public funds meant for the Navy seamen's wages. Cort had assigned his patent rights to his uncle as security against the loan. Both men were ruined and the patent rescinded. It is doubtful that any of the ironmasters allowed themselves to be much troubled by Cort's undeserved misfortune. In any case, they would not have had the time for such unprofitable reflection. English iron was growing at a breakneck pace.

Henry Cort had indeed revolutionized iron-making. The puddling and rolling mills could make as much pure iron in one hour as the previous works could make between sunrise and sunset. Where before the amounts had seemed large, the leading mills were now producing unheard-of quantities and at far lower cost. The great rolling mills were powered by Watt engines, as were the bellows. The steam power freed the mills from their need to sit beside a river, and allowed the ironmasters to migrate to wherever they could find the most advantageous source of coal. The Swansea smelters had already found it. Merthyr Tydfil, a town on the River Taff, 30 miles to the north-east of Swansea, became one of the early centres of the puddling industry. Another was the district in Shropshire where the Darby family's Coalbrookdale Works still reigned supreme.

Yorkshire and central Scotland were still other centres. Each of them benefited from the fortuitous coincidence of iron ore and coal in the same mining field. In the eight years after 1788 English pig iron doubled in output.[41] It had doubled again ten years later. This time it was helped along by that traditional incubator of heavy industry – a war for national survival. William Pitt's impassioned speeches had finally convinced George III and Parliament to stand firm and declare its unflinching opposition to the Revolutionary menace that had thrown its shadow over Europe. In 1793 Britain declared war on the fledgling Republic of France.

NAPOLEON AND HIS AFTERMATH

Although the French Revolution had shaken the ruling classes of Europe it had also found many sympathizers, especially in Britain, where the Whig party was broadly supportive of its aims. The path of revolution is never smooth, and in its immediate aftermath a period of persecution and terror engulfed France. Neighbouring governments looked on with increasing concern, especially when revolutionary groups began to grow within their own borders. The last straw came with the execution of Louis XVI. It was this act that prompted Britain and other nations to declare war. The French got the worst of the fighting until Napoleon Bonaparte took the helm. He led his armies to victory first in Italy and then in Belgium, Holland, Italy and various German states until suddenly France found herself at the head of a European empire. There followed a few years of uneasy peace, during which Napoleon implemented some much-needed administrative and legal reforms. Then in 1804 he was back on the offensive. Once more he found himself at odds with the British. The first real test of strength came in the following year in the seas off Cape Trafalgar, very close to the Straits of Gibraltar. There Admiral Nelson demolished the French and Spanish navies. From then on Napoleon concentrated on developing his Continental System, which had the purpose of sealing off the Continent from trade with Britain.

It had devastating effects on the British economy.[42] Factories stood idle for lack of markets. The banking system was strained to the limit under the inflationary effects of war financing. British manufactured goods were stockpiled in the Mediterranean islands of Cyprus and Malta. At the least sign of a hole in the Continental cordon they were shipped to the mainland to be eagerly purchased by the deprived Italians, Spanish, Germans and even French. In the meantime Napoleon was becoming increasingly stretched to stay atop his Empire. Serious trouble raised its head when the Russians, after five years of alliance with France, switched their allegiance to Britain. In response, Napoleon assembled his forces on the banks of Poland's River Neiman, his purpose to eliminate this developing menace. Instead, the Russian adventure became his downfall. By the time his retreating soldiers reached the safety of Poland once more, the former almost 500,000-strong Grande Armée could call to muster only one-fifth of that number. The triumph over Napoleon strengthened the Tsars, and was one reason why Russia's antique administrative, social, economic and legal system remained in place for another century, until it was at last swept away by the Bolshevik Revolution. In the same way, the Habsburgs' part in defeating the French assisted them to stay in power until their Austro-Hungarian empire was broken up in the political restructuring that accompanied the end of World War I. One equally antique empire was not to be so lucky. Napoleon's overthrow of the Spanish king triggered a wave of Latin American revolutions that was to finish with the Spanish clinging to power in only a few of the islands off the coast of Central America.

The independence movements in the Spanish lands of South America were led by two men quite different in character. The first was Simón Bolívar, a young charismatic and idealistic Spaniard who began his revolutionary career in Venezuela and ended it in Peru. Such was the Liberator's fame that the Republic of Bolivia was named in his honour. José de San Martin was the second. Where Bolívar was dashing and idealistic, San Martin was methodical and practical. He led his forces across the continent from Brazil to Chile and northwards to Peru, where he besieged the capital, Lima. In this process of ridding themselves of the Spanish yoke, the Latin American colonies fractured into a kaleidoscope of republics. Mexico managed to retain its political unity, but it was here that was felt the full ferocity of revolution. The period from 1810 to 1821 is known as the War of Independence. Events began with a priest, Hidalgo, who led hordes of Indians and poorer descendants of the Spanish settlers in destructive and vengeful attacks on the ruling classes and its outward trappings of wealth. The mines were the most obvious of targets and they were ravaged accordingly. Smelter houses were torched, shafts deliberately filled in or flooded and mints were wrecked. Mining activity, suffering not only from the destruction but also from the loss of the government subsidies that had supported it in recent decades, ceased almost completely.[43] Even where the infrastructure was not destroyed it was far too dangerous to risk being caught transporting bullion. The pumps shut down and shafts were left to flood and their timber supports to rot. Hidalgo was captured and executed, but his place was taken by others. Many of the wealthier Spanish – among them the experienced mine operators – were forced into exile.

A similar story could be told of events in Peru, although there the onset of wholesale destruction was delayed by a decade. The Peruvian mining centres of Potosí, Cerro de Pasco – which had recently taken over from Potosí as the largest producer of silver in the Vice-Royalty – and others were all prospering even when revolution was sweeping the continent and San Martin's 'patriot' army was making its slow advance on Lima. While all this commotion took place in the lowlands, the greatest worry of the Peruvian miners was not the patriot armies but the fact that mercury supplies were being disrupted by the fighting in Spain and by the British naval cordon. Such was the serene confidence felt in these remote mountain centres that when Venezuela and Colombia were aflame with revolution, the first steam engines for use in Peru were ordered. It may be noted that the orders were confirmed despite the objections of Boulton and Watt, who insisted that at the high altitudes of the Andes the machines could not properly function. Contrary to Watt's advice, they worked when properly installed with the help of the Cornish engineer Richard Trevithick.[44]

But this sort of modernization was all a little late. When San

Martin's armies arrived, pitched battles were fought around Cerro de Pasco. The skilled workers fled and the steam engines, along with all the other equipment, were wrecked. When the final battle was fought at Junín, a town only a few miles from Cerro de Pasco, the legendary flood of silver from Peru had dried to a trickle. Once the destructive phase of the revolutions had been spent, much of the Latin American mining industry lay in ruins or had been abandoned and left to its fate. At a loss, the revolutionary governments turned to Europe. Yet much of Europe was itself recovering from the devastation of Napoleon's final years. Only in Britain did there exist plenty of unused production capacity, mounds of accumulated capital and an entrepreneurial class that had had its trading and investing instincts thwarted for far too long.

Soon South America was swarming with British merchants. Initially they devoted themselves merely to selling their goods, but inevitably their interest turned to the mines. To the British investor these looked like a great opportunity. After all, the silver and gold of South America was famous, and surely all that was needed were a few modern pumps and some skilled workers and in no time the wrecked shafts would be pumping out profits once more. Pamphleteers egged them on. The result was a mining investment craze.[45] By 1825 hundreds of millions of pounds sterling had been spent on buying mineral leases, purchasing up-to-date equipment and transporting it along with the obligatory Cornish experts to the silver mines of Mexico and the gold fields of Brazil. Money was also poured into railways and all manner of industrial enterprises. Few of the investments ever made a profit. Indeed, the British investments in Mexico failed so resoundingly that for generations afterwards the British in that country retained an unsavoury reputation.

FREE TRADE

After the Battle of Waterloo, Britain was militarily and economically the dominant global power. As its factories grew in size and its goods fell in cost, many realized that the point was nearing where Britain's industrial manufactures could no longer be absorbed by the domestic market. Its continued prosperity would depend on the creation of a dependable export market. This was hampered by the tariff wall that surrounded the island nation. It was an era when tariff protection was routinely used in order to nurture domestic industries. Britain's thriving tinplate industry had got its start after the imposition of a tariff blocked out the German competition.[46] The tinplate story – with its combination of innovation, tariff protection, vigorous expansion and export to the colonies – was one repeated many times over in industrializing England. During the Napoleonic War the tariff barriers had been raised even higher. After the War the merchants of London drafted a petition of protest. If Britain wanted ready access for its exports into South America and Europe, they argued, then it must take the lead in tariff reduction. Their case received support from the intellectual climate of free trade that had been promoted by writers such as Adam Smith and David Ricardo.

William Huskisson, the Secretary of the Board of Trade, agreed. He set about reducing trade barriers and putting the confused structure that remained onto a more rational footing. Out went the lofty tariffs on all manner of items, including copper, lead, zinc and iron.[47] Two years later the Customs Act of 1827 included metalliferous ores among the items that could be imported completely duty-free under 'bond'. This meant that they were duty-free provided they were re-exported after processing. Bonded ores were also exempted from the provisions of the Navigation Act, a centuries-old statute that stipulated that all imports to Britain had to be carried by British ships. With the Huskisson reforms complete, the way was open for Swansea and other British ports to receive the ores of the world. In the years that followed, ships carrying foreign ores began tying-up at the Swansea docks. Chief among them were those from the distant land of Chile.

For most of the colonial period Chile had been a modest producer of gold. The silver mountains of Peru to the north and the Brazilian gold to the east had far outshone the meagre output from this thin sliver of land pressed up against the Pacific Ocean. Independence came early. In 1818 Bernardo O'Higgins became the liberator of Chile. Under his autocratic leadership the country returned to a peaceful condition, and Chilean-born *criollos* (descendants of the Spanish) took up the work that had been abandoned by the fleeing colonists. Copper mines were opened up in the northern province of Atacama.[48] They were mostly small and shallow affairs, occasionally worked by a single extended family. Typically, when the richest of the ores were mined out, the miner moved on to another deposit. Smelting technology was primitive. It was all a far cry from the sophisticated, and by now very large, smelting operations that were clustered around Swansea and Bristol.

O'Higgins's successors allowed political power to drift towards the provinces, and in 1830 an exasperated army, seeing the country proceeding down the path of civil war, overthrew them and installed General Joaquin Prieto. Prieto acted the part of benevolent dictator and brought civil order to Chile. He reversed the regional policies of the Liberals and instituted a strongly centralized system of administration. The result was a political stability that was the envy of South America, and which provided an environment for investment that was sadly lacking in the strife-torn republics to the west and north. Three successive Conservative presidents maintained Prieto's broad policy settings. Trade was considered the engine of growth, and mining was considered the engine of trade. Who better to trade with than the British?

It is here that the British merchant firms enter the picture.[49] They instituted a system whereby they would purchase the copper ore from the mines and transport them to the coastal ports. From there they loaded it onto ships that, in turn, would take the cargo via the stormy southern tip of the continent and on to the British ports of Swansea, Bristol and Liverpool. The journey took three to four months, sometimes six, and the heavy loads made the going tough. It seems a long way to send a smallish shipload of ore. That it was worthwhile can be explained by the tremendous technical and cost advantage offered by the Swansea smelters. For by now they had reached a size and technical sophistication that far surpassed anything available elsewhere in the world. The cost of shipping ore from Chile or Cuba to the south-west corner of Britain was an acceptable price to pay when the alternative was to use a far more costly and wasteful smelting process. With South American ores supplementing the Cornish in the first half of the nineteenth century, Britain in some years produced as much as three-quarters of global copper.

REFORMING THE SYSTEM

Although many of the crucial inventions of the industrial revolution occurred in the years before the Napoleonic War, it was in the years after it that British heavy industry revealed what it really could do. Such was the rapidity of industrial growth in this era that the production capacity of the new factories frequently exceeded the demand for their product. A brutal, unfettered competition became the natural way of things. Into this bedlam were dragooned men, women and children – the foot soldiers and cannon fodder of the industrial revolution. The mines, and in particular the coal mines, were among its bloodiest battlefields.

Coal had been the biggest mining enterprise in Britain since about 1650. The demands of steam power and the coke-based iron mills expanded output so much that when Europe got back to normal after Napoleon, Britain was far and away the largest producer of coal in the world. The mines had pushed deeper into the rich seams that were located in pockets throughout northern England, Scotland and Wales. The working day underground had changed little for centuries. Death was a part of the job. The main hazard was the sudden explosions that occurred when coal gas or dust was suddenly ignited. This could occur from a spark or from the heat of the candles that the miners would stick to their foreheads to afford them some minimal light while they worked. In 1708 an explosion had taken the lives of 69 men and children in North Wales. In a similar incident some years later, 39 had perished. This is to say nothing of the far more frequent occasions when one, two or more men were killed. So frequent did these incidents become that at one time the *Newcastle Journal* was asked not to report them.[50] They were simply tragedies: unfortunate but inevitable. No inquests were ever held into the deaths.

In the mines of Cornwall the death toll was probably not much lower. Here, however, the greatest hazard was the flash flood that occurred when a miner thrust his pick clear through into an old abandoned shaft long since filled with water. The underground deaths were only the tip of an inverted iceberg, for most of the premature deaths were met on the surface. The cause was no mystery. They were the result of sickness and exhaustion brought on by a life of extreme labour, near-starvation wages and constant breathing of coal and quartz dust. The great technical advances of pumping and explosives actually made working conditions worse. As the shafts went deeper, so the end of the day climb to the 'grass' above became ever more arduous. In the deepest of the Cornish shafts, this exhausting climb could take two hours. It was common for men to remain underground for days on end, too exhausted to make the journey up the slippery ladders.[51]

As competition became fiercer, the easiest cost to drive down was the size of the miner's pay packet. Better still, women and children could be used to perform many of the tasks. They were much cheaper to employ and more tractable as employees. Children as young as five worked fourteen-hour days in total darkness. In some collieries, women and young children were employed to drag wagonloads of coal and ore to the shaft along tramways running through tunnels too narrow for a man. The practice became common of apprenticing pauper children to the mines. Only tradition and custom stood in the way of employers doing as they wished. The government preferred to leave such things to work themselves out. While this policy of non-interference must be counted as one of the great strengths of industrial Britain, it showed its weaker side when employment conditions are considered. Emigration became one escape, and the years following the 1820s saw an increasing number of single people and families electing to seek a better life in the United States, Canada, Australia and elsewhere.

In the social upheavals that accompanied the industrial revolution, the moderating influence of custom and tradition was greatly weakened. Society, indeed, was being turned on its head. The fencing of farmland meant larger, more efficient farms, but also less work for many. Those dispossessed were forced out of their small communities and into the large towns, where they found themselves tending the new machinery, their lives now governed by the needs of the machines instead of by the rhythm of the seasons. As more people were pushed into the factories and down the mines, their plight became more obvious to society at large. Voices were raised in protest, but the disasters and abuses continued. One incident of particular horror occurred at the Heaton Colliery.[52] In this instance, the miners were tunnelling into an abandoned shaft when water unexpectedly burst through from the old workings. Seventy-five men and boys were trapped in the tunnels and caverns. Pumps set the task of clearing the flooding took fully nine months to complete their job. When at last the rescuers entered the workings they found a nightmarish scene. Horses, candles and even bark from the wooden props had been eaten. One man had not long been dead. No serious action was taken to prevent a recurrence.

The 1830s was a decade when popular anger over such happenings burst forth in marches, petitions to Parliament and occasional riots. The Chartist movement was formed with the aim of securing legislation to regulate working conditions in the factories. Their methods and goals sometimes looked too revolutionary for comfort. The governing class struck back, sentencing six labourers from Tolpuddle in Dorset to seven years transportation to a penal colony in Australia for the offence of signing up union members. Unionism was illegal being, as it was, a 'combination'. The hypocrisy was all too obvious – one need only point to the Associated Smelters – and the sentence grossly out of proportion. A massive petition succeeded in fetching the men back to England. Political leaders reluctantly began to listen. One of them was to emerge as a champion of the working class. Lord Ashley, or Lord Shaftesbury as he was later known, took up the cudgels against the widespread abuses.[53] His first success came in 1833 with a Bill that limited to ten the daily working hours for people under eighteen. Its other important provision was the creation of a body of factory inspectors. Over time this group of men was to do a great deal in educating the majority and bringing to book the most unscrupulous employers.

The Bill applied only to the textiles factories, and Ashley was later commissioned to undertake an enquiry into the working of the reforms and to widen this review to include other industries. The first of these new reports covered the subject of coal mines. Few of the readers of the report had any first-hand knowledge of conditions underground, so the Committee resorted to the expedient of illustrating the report. One of these illustrations featured a scantily-clad young woman working underground in the company of men.[54] The report created a sensation and led to the Coal Mines Act, which, though it met opposition in the mine owners' stronghold of the House of Lords, was pushed through by the great reforming government of Robert Peel. Women were prohibited from working underground, as were boys under the age of ten. The legislation also created a professional corps of mines inspectors. When it was realized that safety had not been addressed by the legislation, the oversight was remedied by the passage of the Coal Mines Inspection Act. It gave to the inspectors the task of enforcing safety standards down the mine. The Royal School of Mines was founded to train the inspectors in their new duties. The fatal explosions continued, but over time they lessened as standards were improved. No longer could owners neglect their responsibilities with impunity. The first prosecution under the Act occurred in 1856.[55] Britain, Europe's greatest mining nation, had finally followed the centuries-old precedent of their counterparts on the Continental mainland.

In the face of the legislation and new ideas sweeping the country, even the most hidebound and conservative of mine owners could not forever stay impervious to change. New and more effective management practices began slowly to work their way through the industry. The most influential proponent of improved practices was the Norwich-born John Taylor. He got his start in the closed world of Cornish mining with the assistance of a Devon-based family, the Martineaus.[56] They were rich friends of his parents, and owned the Wheal Friendly copper mine near Tavistock on Dartmoor. Either the Martineau family was prepared to take nepotism a very long way or it saw something unique in Taylor. For his utter lack of experience was no barrier to him being given the job of manager of the Wheal Friendly at the age of nineteen. Taylor did not let down his patrons. Fourteen years later, when he departed the district for London, he was one of the more prominent mining experts in the country. His success was based on his conviction that a mining concern should be managed and developed in a rational fashion with an eye to the long term. This was a radical departure from the prevailing attitude according to which a shareholder grabbed his yearly profits with barely a thought for the next year. Taylor also relentlessly pursued cost reductions and economic mining methods. He was a firm and outspoken advocate of the tributing system, considering it the best means of aligning the interests of a mine's shareholders and its employees.

The origins of tributing date from the time when the Cornish mines moved underground.[57] The small alluvial and hardrock workings of yesteryear had been worked by teams of casual labourers led by the owner of the claim. He was called the Captain. The teams were free to disband at harvest time and other periods when the working of the pits was not suitable. Clearly, such a system would not suffice in a mine where expensive capital equipment had already been installed. Thus, the question for the mine owners became one of how to shape the new workforce and at the same time minimize their wages. The answer that evolved came to be known as the tributing system. In essence it was a system in which teams of miners, each led by a Captain, would bid against each other for the right to a certain block of ore. The amount of the successful bid was payable to the owners. The Captain and his team would then proceed to mine their block of ore, and at the end of the contract period they had then the right to sell it to the smelters for the price the smelters offered them.

Tributing had some superficial advantages for the working man. First, it was built on the traditional unit of the Captain and his team. Second, it offered to the miners the chance to strike it rich. If, for example, their block of ore turned out to be unexpectedly rich, then they would receive a windfall profit. The system, it was said, suited the innate Cornish desire for independence, and indeed the working hours and work methods of the tributers were largely determined by the Captain. Unfortunately, the system suited far more the innate desire of the mine owners for profit without risk. As in most casinos, the odds were stacked in favour of the house. The system saw to it that the teams were pitted against each other in bidding for the blocks with the result that the tributer took all the risk and received few of the rewards. The result, as it had been back in Tudor times, was that the working men very often received barely a living wage. To make matters worse, they were paid only once during their contract – at the end. During the term of the contract they frequently had to receive advances from the mine owners, advances they often could not repay. They became increasingly tied to the mine. It is surprising that Taylor could support such a system at a time when many were raising their voices against it. One can only suppose that he was in the habit of paying his workers higher rates than was typical.

Taylor's achievements in Devon had been noticed further afield, and from his London base he was invited to advise on, and eventually assume, the management of all manner of metal mining enterprises throughout Britain. In this way he became involved in the lead fields of Staffordshire and Yorkshire and in the copper mines of Camborne and Redruth. Later he was to head the British advance into the lead-fields of southern Spain.[58] In 1819 he assumed control of an abandoned mine near Redruth, the Consolidated, once widely known for its rich ores but now a drowned shell. A decade later the Consolidated was the largest copper concern in the world. Such was its success that the name 'Consols' long remained one that unscrupulous promoters would adopt to give their prospectus a touch of magic. Taylor brought to the Consolidated not only many technical improvements but a new style of management, one that challenged the cosy links and practices of Cornwall. In this he was strongly backed by the new-founded *Mining Journal*. He instituted a system of competitive tender for all contracts and began to source his finance from bankers in the City of London.

The change in financing was no minor matter. In Taylor's day, the skill and honesty of the management of mining ventures were so dubious that such investments were shunned by the mainstream financial community. The Stock Exchange refused to deal in mining shares.[59] Taylor's competence and integrity allowed him to begin to change this. His work left two important legacies. First, he helped lay the foundations for mining ventures to be financed through the London Stock Market, a development that was to be an indispensable ingredient in the explosive expansion of the 1880s and 1890s. Second, he established the model whereby a single team of professional managers could take on the conduct of numerous mines, geographically dispersed. This model, which made much use of the 'consulting engineer', was to be widely adopted at the end of the century, when the pace of developments began to outstrip the availability of skilled professionals.

A NEW SCIENCE

As John Taylor was introducing a more systematic approach to the activity of mining, systematic observation was being applied to the study of rocks. It was not a completely new field. We have seen that a systematic approach to the study of rocks can be found at least as far back as Agricola. It was he who wrote a treatise in which different rocks were categorized in terms of properties like hardness, ease of smelting and so on. However, if we are to look for a period during which the study of rocks moved from the status of a compendium of observations to a science in which those observations were coherently explained by a set of principles that had been tested by vigorous debate, then we may fairly settle on the second half of the eighteenth century.

By the beginning of this period a theory claiming to explain the pattern of the earth's rock structure had been developed. It held that the geological strata so clearly visible in cliff faces were the result of successive layers of rock being deposited out of a mineral-heavy ocean that had once covered the entire surface of the earth. The leading proponent of the theory was a German professor, Gottlob Werner, who was teaching at the Freiberg School of Mines.[60] By virtue of their reliance on the theory of oceanic deposition, Werner and his supporters were known as Neptunists. While the Neptunists offered an explanation of the existence of rock strata, they could do little to explain the existence and distribution of actual physical land forms such as broad plains and jagged mountain ranges, and in time there developed a rival camp. The Plutonists, led by James Hutton, claimed that in contrast to the once-only deposition event propounded by the Neptunists, the earth was in fact continually changing its geological structure. This was, the theory went, because it is a heat-generating, dynamic body. According to the Plutonists a typical geological cycle began with the formation of mountain peaks, when the outer crust of the earth crumpled under pressure from the heated mass below. Erosion then gradually wore down the mountains and in the process formed additional layers of sedimentary strata. Then the cycle began again.

An unusually strong rancour characterized the debate between the two schools. The argument was still raging without apparent hope of resolution when a self-educated English surveyor by the name of William Smith introduced amid the clamour the calming influence of systematic data gathering. He set about examining the rock of England. He carefully catalogued the strata and the fossils contained within them, and after some years of research could draw from his labours the conclusion that each different stratum contained its own characteristic fossilized life-forms.[61] Smith also concluded that these stratified fossil groups showed evidence of a progression from simple to complex as one followed them through the layers. His work did little to resolve the Neptunist–Plutonist controversy and, alas, triggered a second. For the question naturally arose as to what could explain certain discontinuities in the fossil record. Two answers offered themselves. The discontinuities could be due merely to absence of evidence, or they could be positive evidence that over the course of the ages the earth had experienced a number of catastrophic events, each of which had abruptly changed the mix of animal and plant life. Tempers, already frayed from decades of conflict, flared again and once more two schools – the Uniformitarians and the Catastrophists – took up their places on opposing barricades. The Uniformitarians held that the rate of erosion and new peak formation had been constant over the ages, and thus any sudden shift in the fossil record must of necessity be simply the result of an inadequate fossil record. The Catastrophists, on the other hand, maintained that occasional monumental upheavals in the earth's crust had extinguished life in entire regions.

The bitterness that attended both these controversies may be partly explained by the fact that they touched a raw nerve in the body politic. For science was being brought to bear on matters long the preserve of biblical scholars. Only 100 years before Hutton had propounded his thesis, the Anglican Bishop Ussher had added up the ages of all the characters mentioned in the Book of Genesis and finished by computing that the birth of the earth must therefore have occurred in 4004 BC. (A further refinement of his work found more precisely that the time of the Creation was at 9 am on 26 October of that year!) Whilst the Neptunist school could possibly be shoe-horned into conformance with this estimate, there was no possibility that the same could be done with the Plutonist view. Thus the geological theories of the eighteenth century placed more at stake than merely the way in which rocks had come into being. The principles of the argument were debated by well-informed people throughout society. Charles Lyell's book, *Principles of Geology*, was read aboard the *Beagle* by Charles Darwin and is credited as having been an important influence on the final shape of his theory of evolution. Lyell was a leading Uniformitarian. Had Darwin taken with him a volume written from the Catastrophist point of view, his eventual theory may well have taken on a different aspect.

What did any of this theorizing and controversy have to tell the seeker of minerals? In short, nothing. It would be a long time before the macro-theories of geology would yield any practical guidance for the mining man. Of more use was the work of William Smith and his followers. Having established his observations on fossil progression, Smith had set about collecting his detailed observations into a series of maps of the surface geological formations of the British Isles. These were eventually published in 1815 in a book entitled the *Geologic Map of England and Wales with part of Scotland*. It was the world's first thorough geological survey, and provided the inspiration for a host of others. The maps had practical applications. Smith was frequently engaged in canal construction projects; indeed it was from his surveying and engineering observations during the course of this work that came the inspiration for his geological studies. His maps provided a valuable guide in the planning and construction of the capital-hungry canals.[62] It soon became apparent that a knowledge of the rock strata could provide directions, however unreliable, to the location of coal seams. Thus the survey maps also came to be sought after by landowners. For the demand for coal was such that every county squire was on the lookout for a payable deposit. From that point the application of the surveys to the search for metal ores was only a matter of time and experience. The field of mine geology had been born.

BRITISH STEEL

Watt's pump patent expired in 1800. By then Boulton and Watt had long ceased being regarded as saviours and instead were viewed as parasites by the majority of miners. The annual dues payable to them were frequently withheld in the hope that someone would successfully challenge the validity of the patent. In many instances mineowners had installed engines of Cornish design but based on Watt's famous separate condenser. Some of them refused to pay royalties to Boulton and Watt. Initially the two, now wealthy enough, had let such challenges go, but later their attitude changed. Watt took the view that to let them go was to admit that his patent was somehow fraudulent. His life's work and reputation depended on the partnership taking a tough line. Accordingly, they began to sue the mineowners and the engineers who supplied the engines. On the whole they were successful, and they certainly succeeded in stifling the incentive for home-grown innovation.[63] Such was their unpopularity that when Watt's patent expired, the firm abandoned Cornwall. The Cornish engineers were now free to bring their own experience and creativity to the subject of pumping. Of the many who took up the challenge the most successful was Richard Trevithick.

Trevithick was an inventor of outstanding ability. The son of a mine manager, he had managed a small mine himself before turning his hand to engineering. Among his first inventions was a pump that allowed the use of steam at a far higher pressure than Watt had ever dared. The result was a patent for a pump that was smaller, more powerful and used less fuel. Other engineers added their own innovations, and before long steam-powered pumping was viewed as a Cornish technology. The improvements were driven by a spirit of intense but friendly rivalry between mines as to which could achieve the greatest pumping efficiency. Record performances occurred year after year, and Cornish pumps and engineers were soon in demand the world over.

Trevithick followed up his pumping advances by constructing the first steam-powered rock drill. Next he built the first steam-driven locomotive, which he drove up a hill in Camborne. He drove a second model through the streets of London, and the following year used the same engine to haul 70 men and a load of iron along ten miles of tramway in a South Wales coalfield.[64] Few seem to have grasped the significance of this epoch-making event. The steam locomotive failed as a commercial proposition and horse-drawn wagons remained in vogue. Later in life Trevithick travelled to the Andes to oversee the installation of his pumps at Cerro de Pasco. He remained there many years, living through the tumult of the revolutions, before finally returning penniless to Cornwall. He died in poverty and was buried in an unmarked grave. It was a fate that may well have befallen James Watt had he not teamed up with one of the best businessmen of his time.

The idea of a steam-driven locomotive for hauling heavy loads of coal and ore was too powerful to be lost for long. Working independently, George Stephenson and William Hedley both unveiled functional steam locomotives in 1813, and the next year Stephenson had his model working in the Killingworth Colliery, hauling 30 tonnes of coal at the rate of four miles per hour.[65] The next step along the path to steam-powered transport occurred nearby. The Stockton and Darlington railway was originally to have been a tramway of the horse-drawn variety, and the line was more than half built when the promoters paid a visit to the Killingworth Colliery to observe Stephenson's steam locomotive in action. They were hooked. So it was that on the grand opening day of the Stockton to Darlington line, the crowd was treated not to a display of liveried horses elegantly drawing a string of wagons but instead witnessed a display of brute machine power. George Stephenson's single locomotive was able to pull 80 tonnes of coal at more than ten miles per hour. From that day onwards horse-drawn tramways were on the way to extinction. Then came the opening of the Liverpool to Manchester line. Its importance was to show that railways need not be solely employed in the hauling of bulk minerals. The Liverpool–Manchester line was, instead, a stand-alone passenger and goods service running between two major cities.[66] A new, fast and versatile form of all-weather transport had arrived.

With this Britain was catapulted into the railway age. For the next twenty years the country found itself in the midst of a mania as investors clamoured for shares in the great passenger lines then being laid. The laying of the national railway network created, of course, a huge demand for iron. The rails had perforce to be made of puddled iron; anything else was too brittle and would crack under the weight of the locomotives. So it was that between the end of the Napoleonic War and the middle of the century, production from the puddling works rose sevenfold. This meant that since the year that Cort's patent had been granted, the output from puddling in Britain had doubled every ten years. Iron works got bigger and their number multiplied. The iron-making process was, however, still laborious and costly and the rails it produced were too soft for their task – they had to be replaced every year or even sooner. So when at last a new process was announced, it was seized on by ironmasters throughout the country.

In August 1856 Henry Bessemer read a paper to the Mechanical Section of the British Association.[67] In it he detailed his recent successes in devising a process for making iron as good as the puddled variety simply by blowing a stream of air through the molten pig iron. It was stirring stuff for a hard-pressed ironmaster. Three days after it was delivered, *The Times* printed Bessemer's paper in full. Within two weeks the Dowlais Company, a large and long-established South Wales iron works, had agreed to license the process, and in the ensuing months a stream of British and Continental ironmasters followed the Dowlais example. They soon regretted it. Not one of the licensees was able to make acceptable quality metal from the process. Bessemer, a newcomer to the industry, had dared to claim a major advance and had been shown to be an imposter or at best a 'wild enthusiast'. Facing the public scorn of the disappointed ironmasters must have been a tough test of his character. For his new process was not the fruit of long experience and observation. Rather, it was the result of eighteen months experimentation, the latest in a long line of inventions.

Bessemer was not a metallurgist but a remarkably gifted inventor who had produced advances in fields as diverse as document stamping, the manufacture of lead pencils and sugar-cane crushing. His attention had been drawn to iron-making through yet another of his inventions. In 1854 he had devised a new method of rotating projectiles. Thinking that this would be of some use to the war effort then being mounted in Crimea against the Russians, he offered his latest brainchild to the War Office. On being rebuffed he took it to Britain's wartime ally, France, where the reception was far more amicable. While there he discovered that the existing cast-iron field guns were not strong enough to sustain his new projectiles. It was then that, in his own words, 'the object I set before myself was to produce a metal having characteristics of wrought iron or steel, and yet capable of being run into a mould or ingot'.[68] The metal that emerged from his new process was, in fact, steel. Like Cort's process before him, Bessemer's process would change the face of iron-making.[69] No longer would there be iron works, after Bessemer there would be steel mills.

It was a long and painful birth. First there was the question of why the process had failed when applied in other iron works. Two years of painstaking backtracking were required before he realized that the level of the element phosphorus in the iron ore was the culprit. In his initial experiments he had used Swedish ores, famous for their purity. Those who had subsequently tried his process had used English ores, many of which contained high levels of phosphorus. From then on Bessemer would use only ores with low levels of that element. Robert Mushet found that the quality of Bessemer steel was further improved by the addition after the air-blowing of a quantity of manganese.[70] Once the technical problem-solving was successfully accomplished there remained the problem of gaining acceptance. Among the iron-making community Bessemer had by now a poor reputation. Even when he had built a steel mill of his own and was successfully producing steel at a price cheaper by a third than anyone else, still there were few prepared to follow him. There was more to their wariness than a 'once bitten, twice shy' reaction. The process was, in fact, very difficult to control properly. Pumping air through molten iron resulted in a rapid and violent reaction in which the outcome was unpredictable. So while most of the batches of Bessemer steel were good, every now and then there was one in which the steel was brittle and liable to fracture. When offered steel for his rails and locomotive boilers, the chief engineer of the London & North Western Railway had responded: 'Mr Bessemer, do you wish to see me tried for manslaughter?'[71] This summed up the view even of those who were well disposed to Bessemer.

Time has a way of overcoming such obstacles, and eventually more ironmasters tried and then adopted the new process. Steel had too many advantages. It could be made at lower cost, allowed higher production and was much harder than the metal produced via puddling. When at last the Siemens brothers managed to tame the violence of air on molten iron, the steel age was set to begin. As for the railway, it did not greatly transform the mining industry of the British Isles. Rail certainly helped to reduce its transport costs, helped it expand and over time loosened the geographical constraints on the location of iron mills, copper works and other industrial complexes dependent on large quantities of coal and ores. But it was to be in other lands, where mountainous terrain and vast distances made water-based transport unfeasible, that railways were to have their greatest impact on the development of mining. Chile was one such land.

As its copper output multiplied, Chile had steadily grown in importance as a supplier to Swansea. The growth was facilitated by railways. The first railway of any size to be constructed in Latin America was built in 1851 between the Pacific port of Caldera and Copiapó, the main centre of Norte Chico, as the region was called.[72] A group of wealthy miners and traders provided the finance, and the design and construction was undertaken by William Wheelwright, an American engineer who had earned his reputation by organizing the first paddle-steamer shuttle along the Chilean coast. Thanks to Wheelwright's choice of terminus, Caldera was transformed from a sleepy harbour to a major port. Shipping increased four-fold in five years. In contrast, the former main port of Puerto Viejo was forgotten. Such was the power of the railway. Soon the Caldera–Copiapó line had pushed on through the northern mining district, eventually forming a rough semicircle through the richest copper and silver areas. A second rail system ran to the south. The effect was to create two catchment areas for the ore of Norte Chico and drain it away to the coast. As these rivers of copper became torrents, the country grew wealthy.

Santiago, the capital, and Valparaíso became known for their European air and for the abundance of their millionaires. Yet the wealth remained concentrated in the hands of a few, mostly copper merchants and bankers. The country never developed an industrial base to rival Swansea. Instead, throughout its first copper era, Chile remained dependent on foreigners to process its mineral wealth. Foreigners, too, managed the trade. Many blamed the government for these failings. Its free trade policies, they claimed, ensured that homegrown enterprise died before it could ripen. Furthermore, the miners complained, the government treated the industry as if it were simply a source of revenue. They had a point. The export duties on copper and milled flour were among the main sources of government income. Aggravating matters was the fact that government expenditure was concentrated in the cities of central Chile, far from the northern provinces where the mines were located. Discontent festered in the outlying regions and flared into

4 Chile, 1840

rebellion. Two armed uprisings in the north were actively supported by the wealthy miners as well as the poor.[73] The second, in 1859, saw mineworkers group together in an organized militia and establish a liberated zone. The central government managed to crush both revolts but could not eliminate their causes. The copper question continued to fester even as production grew.

Fed as they were on a rich diet of Chilean and Cornish ores, as well as on liberal side servings of the Australian and Cuban variety, the years between 1840 and 1860 were the period when the Swansea smelters were at the height of their global dominance. The city, polluted and grimy, was the hub of a sophisticated and extensive maritime trade. The spacious port was full of ships. Sometimes a clutch of them would arrive all at once as if in convoy, then there might be none for several weeks or even months. The feast and famine nature of these arrivals played havoc with the ore price. One constant theme of business history is the unfailing ingenuity of merchants in coping with such difficulties, and the Swansea ore traders were no exception. If a ship's cargo, they realized, could be assured of a buyer and a price before it left its port, then the boom and bust nature of the trade would lessen. And so the practice developed of selling cargoes 'forward'.[74] To better manage this increasingly complex trade, in 1876 the London Metal Exchange was established. The Exchange also dealt in the tin shipped to British ports from the straits of Malaya.

Tin had been dug from south-east Asian alluvial fields for many centuries. The deposits lay in a line running south from southern China, through the Malaysian peninsular and ending in the small islands of Bangka and Billiton off the coast of Sumatra. For the most part, the tin had gone to China and India, where it was highly popular when beaten into foil to be burnt as an offering. It was also used in making shiny ceremonial shoes. Since the 1830s the Malayan fields had come to be dominated by Chinese fleeing from the persecution of the Manchu dynasty. Their improved methods of organization had resulted in output expanding until production rivalled Cornwall's. The Cornish shafts were worked at considerably higher cost and many shut down, unable to compete.

Digging the Malayan alluvials was a hard, brutal and often fatal occupation. Chinese secret societies ruled the fields with an iron fist.[75] The backbreaking labour was punctuated by battles with Malays and bouts of drinking and gambling that could easily end in bloodshed. Political violence was common. Occasionally during an extended fracas, an appeal would be made from one of the contending parties for the British, sitting in Malacca, to intervene. They consistently refused. This stance was finally changed when serious warfare broke out between rival Malay camps in the states of Selangor and Perak. The conflict spread and the secret societies took sides. Massacres occurred on the tin-fields and production was severely disrupted. Calls were made for British intervention, and in 1874 they went in. It was over in months, resolved through a classic piece of gunboat diplomacy.[76]

The episode had two outcomes. First, it brought Malaya into

the British Empire. Second, it brought the shipping of Malayan tin to a near standstill, kicking off the biggest boom in tin mining that Cornwall had ever seen. Tin overtook copper to become once more the main metal of the region. The next two years saw prices rise to levels that had not been seen since the years of war against Napoleon. Like wild flowers after a desert rain, the abandoned mines sprang into life. New ones were started up. Pumping engines long disused were belching their smoke once more. Old and deep mines – Dolcoath, Carn Brea and Great Wheal Vor – worked overtime to send their ore to the surface from depths as great as half a mile. They were joined by scores of hopefuls that were only just beginning to scratch the surface. Tin mania swept the country, although it was noted that the most avid buyers of mining scrip were the least informed of investors. Many of these would have shared the simple sentiment of the purser of Wheal Owles, who in a public meeting had 'scorned the idea of a Cornishman being beaten in the production of tin by a Coolie or a Chinaman'.[77] It was simply not in the natural order of things. Much more in the way of the natural order were the speculators who preyed on the gullible. Great profits were there to be made by the nimble, the unscrupulous and the well-connected.

The decades of hardship and the recent years of despair had so built up that the upswing in prices had an exaggeratedly euphoric effect on the county. It was, of course, never going to last. The high prices had sparked worldwide interest in tin. Alluvial deposits were discovered in Australia, and wool clippers laden with tin ore as ballast began arriving in British ports. When order was restored to the Malaysian peninsula, tin from the East was back with a vengeance. The bubble burst. Prices collapsed and share prices went with them. The situation never recovered, and Cornish tin was finished as a force on the world stage. Cornish copper was also in the final stages of its long career. It now came from too far underground to have any chance of competing with the cheap and plentiful ores of Chile, Spain and the United States. With its demise the great era of European mining was at an end. The next 100 years would see global production dominated by new continents that were only beginning to yield up their riches.

THREE

New Frontiers

THE FRONTIER

In contrast to the speed of the conquest of South America, the European invasion of North America was sluggish indeed. Nearly 300 years after the first settlers established themselves along the Atlantic coast, the native peoples still held sway over much of the continent. By far the majority of the new arrivals were concentrated in the young republic of the United States. In Virginia, the Carolinas and Georgia, cotton and tobacco were grown and harvested on large slave plantations. Eli Whitney had invented the cotton gin in 1792, and the cotton fields were flourishing as never before. The northern States were home to great fishing fleets and a range of manufacturing enterprises. A little copper was mined and smelted in Vermont. Iron was the main metallurgical industry. Hundreds of small furnaces fed with locally produced iron ore and charcoal had once flourished to the extent that iron had been exported in considerable quantities. The growth of British iron had put an end to American exports, but the domestic industry remained strong.[1] The same could not be said of manufacturing as a whole. The United States and the British colonies to its north both relied on Europe for most of their manufactured goods. In return they shipped fish, raw cotton, wood and furs back across the Atlantic.

The citizens of the United States and Canada were not the only Europeans on the continent. The Vice Royalty of New Spain – soon to be the Republic of Mexico – laid claim to a great L-shaped expanse that encompassed much of the western and southern sides of the continent. It covered what is now California and all the arid south-east as far as Texas.[2] For centuries the Spanish had occupied this land without realizing the great mineral wealth it held. A few hundred miles from their

Californian coastal settlements lay a treasure trove of gold barely concealed in the streams and gorges of the Sierra Nevada mountains. The gold and silver of Arizona and Nevada lay similarly untouched. Why were these riches not found when the Spanish were ransacking the length and breadth of Mexico and much of South America for the very same precious metals? It was partly a matter of chance. In the early years of Spanish occupation several prospecting parties had ventured north from New Spain via Arizona and Texas as far as Florida in the east and Kansas in the north. They had found nothing. From then on the Spanish population had remained sparse. Those who did live in the region were mostly Franciscan and Dominican friars and their assistants. The friars were keen to avoid the abuses perpetrated by their countrymen in the lands to the south. They knew a gold strike spelled doom for the Indians. For this reason, when occasionally an Indian would bring to them an acorn cup full of gold, his offering would be spurned in the hope that it would discourage him. In Mexico or Peru that same Indian would have been knocked over in the rush.

The vast territory of Louisiana occupied the middle ground. It lay between the Rockies in the west and the Mississippi in the east. Here French adventurers conducted trapping operations over the plains and mountains, the home of hundreds of Indian tribes. The French had the happy knack of being able to establish good relations with the Indians. With their permission and protection lead was mined in several places along the Mississippi River, especially in Illinois and Wisconsin.[3] It was mostly used to make pellets for the trappers' rifles or was floated downriver to the main French settlement of New Orleans. In 1804, needing money to fund his campaigns, Napoleon sold Louisiana to the US government for a bargain-basement price. With the Spanish asleep over in the west and the Indians unable to offer a match for European firearms, it must have seemed to the ambitious and energetic people of the United States that the continent was now theirs for the taking. In their thousands they started moving westwards. And so the frontier was born.

The westward movement of the American people has been compared by many historians to the eastward migration that occurred in Europe during the Gothic age. The motivations were similar. They included hunger for land and a general restlessness and dissatisfaction with life as it was. Frontier life tends to be romanticized. In truth there were among the pioneers few philanthropists. 'GAIN! GAIN! GAIN! is the beginning, the middle and the end, the alpha and omega of the founders of American towns', wrote one English traveller.[4] If the Indian stood in the way he was pushed aside – slaughtered if necessary. Land grabs, sharp dealing and violence were typical of the way affairs were conducted out on the frontier. By 1840, all the land up to the Mississippi had been incorporated into the United States and was steadily being filled with settlers. Mineral prospecting was generally not uppermost in the mind of those who crossed the coastal ranges and drove towards the Mississippi, but discoveries were inevitable with so many new pairs of eyes observing the land for the first time. A gold strike was made in Georgia and was followed by others in Alabama, Virginia and the Carolinas.[5] They were small affairs.

The lead mines of Wisconsin were expanded and others were developed in Missouri. It was all for local consumption. The cost of transport meant that it was cheaper to ship lead from Europe to the main cities of the eastern seaboard than it was to cart it there all the way from the Mid-West. In the north, bordering the Great Lakes, there had been persistent rumours of country rich in copper. The local Indians were said to keep large lumps of native copper as mascots. For long, however, the region was too remote to seriously interest anyone but trappers. But times were changing. As the westward expansion continued, the state of Michigan was created. Soon after, Douglass Houghton, Michigan's first Chief Geologist, commenced a geological survey of the state. He identified the Keweenaw Peninsula, a narrow finger of land poking into Lake Superior, as the copper country of rumour.[6] Miners from Wisconsin rushed to the area. They were followed by a small army from the states further east. The ore had to be shipped through the Great Lakes to Detroit, from where it was railed to Baltimore. Prices were low and the transport costs high. Chilean ore was cheaper to the East coast smelter owners, and so the copper of Michigan lay substantially undisturbed in the forests. Such was the mining scene on the North American continent as the mid-century approached.

It was an era when the Europeans everywhere began to move into the 'unoccupied' lands of the world. In Australia, the British colonists crossed the rugged mountain range that separated Sydney Cove from the plains inland. They also began to establish other settlements around the coast. On the Brisbane River, far to the north of Sydney, a convict settlement was created. The new colony of South Australia was founded, and over in the west of the continent the town of Perth was built on the sandy banks of the Swan River. As in the United States, it was inevitable that the migrations and new settlements would uncover fresh mineral fields. One such find was the copper of South Australia. The first of several major discoveries was made north of the major port town of Adelaide by a shepherd, and soon the rich copper ores of the colony were being dragged by bullock cart to port. From there they would be loaded as ballast into wool clippers bound for England.[7] By acting as ballast the transport costs to Swansea from South Australia were lower than for the voyage from Chile. Assisted by this advantage, by mid-century South Australian ores were an important and growing part of the Swansea global network.

In Africa, too, the territory under European occupation was expanded. Southern Africa was the scene of the greatest activity. During the Napoleonic War the British had seized the former Dutch East India Company outpost of Cape Town. Conflict arose between the British and the local Boer community, and the Boers had set off on their Great Trek to the north. It was a migration that would eventually occupy centre-stage in one of the more drama-filled episodes in the history of mining. But the diamonds and gold of southern Africa belong to a later period. It was the gold of other continents that would be the first riches to be won from the new frontiers.

SIBERIA

There are parallels between the experiences of the American frontier and the Russian, but the differences are equally great. Where the United States was founded on democracy and an aggressive sense of independence, Russia was a nation dominated by the god-like figure of the Tsar and the pervasive influence of the Orthodox Church, with its ritual and conservatism. Ignorant and drunken were words used by many visitors to describe their aristocratic hosts. The peasants were bound to the land, their lives forfeit to their master. That Russia was the laggard of Europe was no new development. It was a problem first faced by Peter the Great. Coming to the throne in 1697, it had been his single-minded mission to Westernize his insular and backward country.[8] During his blood-soaked reign, he attempted to bludgeon and terrorize his subjects into adopting Western manners and habits in the hope that they would eventually exhibit the technical skills of the Europeans. He probably succeeded more in traumatizing his countrymen than in changing them. Mining and smelting, however, did grow with Peter's encouragement and on the back of burgeoning European demand. Let us take a moment to examine this first flowering of Russian metallurgy.

Peter's interest in metals was driven by the needs of his armies. For many years Russian supplies of iron and copper had been purchased from Sweden, which, as we have seen, was at that time producing two-thirds of Europe's copper and was the source of the Continent's best quality iron. Those supplies had been cut-off when Gustavas Adolphus invaded the westernmost lands of the Russian Empire in an attempt to include them in his version of a Swedish commonwealth. Deprived of the metals essential to equipping his fighting men, Peter determined that, as in so many other areas of his economy, the time had finally come to be self-sufficient in such materials so that his giant but weak country could not again be held to ransom by its smaller neighbours. This concern seems to have been the reason why he invited to his court a young ironmaster named Nikita Demidov.[9] The meeting must have been a success, because in return for a guaranteed annual supply of iron at a low price, Peter granted Demidov the exclusive right to produce iron in the Magnit Mountains region of the Urals. As part of the bargain the Tsar also funded the building there of a large state-of-the-art iron works. He had chosen the right man. By 1721 the Demidov industrial empire was producing over half of Russia's iron and was opening the first of many copper works. It had become a central pillar of the state.

On Nikita's death the responsibility for the enterprise fell to his son Akinfii, in whose hands the growth continued until the family's wealth earned them the sobriquet of Lords of the Urals. Capable as he was, Akinfii seems sometimes to have allowed his greed to get the better of his judgement. His most serious error in this regard came after the discovery by trappers in the distant Altai Mountains (near the present border with China) of an ancient and long-abandoned copper mine. The find was brought to the attention of Akinfii, who duly sought and received Kremlin permission to construct there a smelting works. It came to be known as Nerchinsk. Soon silver was found to accompany the copper. When a richer silver–copper deposit was later found nearby, the younger Demidov decided to conceal this new bonanza from the Crown. He got away with it for nearly ten years, an outcome that must owe a lot to the fact that the works were situated 1,000 miles from the Demidov centre of operations in the Urals, and these, in turn, lay nearly 1,000 miles from Moscow. At last Akinfii's dishonesty was uncovered, and from then, unsurprisingly, the family fortunes declined. Yet others had seen what could be done, and Russian production of copper and iron continued to increase until for several decades the country was a world leader in the production of those two metals. In France, Russian copper was more common than the Cornish variety. Russian iron was a staple in many British foundries.

Nerchinsk lay on the frontier, and attacks by local tribesmen were regular events. The works were fortified and equipped with large supplies of guns, cannon and ammunition. Special precautions had to be made for the delivery of lead for the liquation works, which was dragged over rough tracks from mines in the Ukraine, 3,000 miles to the west. The danger did nothing to slow its expansion, and eventually the complex would earn for itself the title of 'the Russian Potosí'.[10] Thanks to Nerchinsk, for a time Russia was nearly self-sufficient in the precious metals needed for its coinage. Given this thriving industry of iron, copper and silver, how was it that when Tsar Nicholas came to power in 1825 he found his empire to be but a minor player on the global metals stage?

The causes were several. The Tsars had always had difficulty in trusting powerful entrepreneurs, and the experience of Demidov's duplicity underscored this. The incident was not forgotten, and over time many of the smelters and mines were brought under control of the state, where lax administration had weakened them. More seriously, Peter the Great's legacy had left dangerous weaknesses in the social fabric. Peasants had borne the brunt of the taxation that had been raised to fund his grandiose building projects, such as the construction of St Petersburg. Later Tsars did little enough to repair the damage. In 1783, their patience at an end, hordes of peasants rose up throughout the southern Caucasus. It was called the Pugachev rebellion. The rich and their assets were destroyed in a frenzy of burning and pillaging that was halted only with difficulty by the loyal Cossack cavalry. Then there was the fact that the Cort process and the industrialized combination of Cornwall and Swansea had taken Britain to a far stronger position as an international competitor. Taken together, the Pugachev rebellion, the mass production from British smelters and an unenergetic response to these challenges were enough to drive Russian metallurgical industry into hibernation. It was to lie dormant for 100 years before another tyrannical modernizer, Joseph Stalin, forced it once more into the light of day.

With the passing of Russia's first period of vigorous eastward expansion, the occupation of Siberia assumed a more leisurely pace. It was still a mostly unmapped and dangerous wilderness when, in 1832, a merchant doing his rounds near the remote Yenisei river basin happened upon a rich alluvial goldfield.[11] This soon led to even bigger discoveries nearby. Thus began the first of the great modern gold rushes. The gold had been found in areas guarded by thick forest that became impenetrable in places. The ground was marshy and the air full of mosquitoes in summer. In winter the area was uninhabitable. Into this unwelcoming environment flocked fortune hunters from all over Siberia. The hardships of the journey were such that many died on the way, while those who got there found that the government had beaten them to it. Instead of the near anarchy that was typical of a new goldfield, order existed in Yenisei from the start. Determined to establish its authority, the government had adopted a policy of granting large areas to merchants and others who could gain the favour of officialdom. Most of the ground was held in the hands of a few. A prospector could not simply arrive at the field, stake a claim and start panning. The new arrival usually found himself employed as a labourer working a claim belonging to one of the big mine owners. His pay was low, and vodka had been banned. Secret police roamed the area, stamping out trouble before it brewed. Then in September, as winter approached, the mining stopped and the entire population headed for home. One wonders what sort of welcome they would have received as many, perhaps most, of the vodka-deprived miners had drunk their earnings well before arriving in their home towns.

CALIFORNIA

As the American frontier moved west, the pioneer gaze turned to the huge tracts of territory held by the Republic of Mexico. First to attract them was the province of Texas, and soon it was home to a sizeable population of American settlers. They and their autocratic hosts did not mix. Clashes followed and the settlers proclaimed an independent government. It was an act that prompted swift retaliation. A Mexican army crushed the rebellious settlers at the Alamo, a palisaded fort not far from present-day Houston. For the Mexicans it was to be a short-lived victory. The settlers fought back and won, and in 1837 their new-founded Lone Star Republic won the recognition of the United States. Eight years later it was formally admitted as a state of the Union. It mattered little that Mexico strenuously protested this annexation of what they considered their stolen territory. Even they must have known that such protest was no more than a futile attempt to turn back the tide of history. War followed, which ended, predictably, with total Mexican defeat, and on 2 February 1848 Mexico ceded to the United States a great swathe of territory stretching from New Mexico to California.[12] The first discovery of gold in California's Sierra Nevada mountain range was made a mere nine days before this transfer of territory. For 300 years of Spanish occupation the mountains had kept their secret. Now, on the eve of a new regime, they revealed their treasure to the world.

An understanding of the gold rush that followed is helped by an appreciation of the geography of the region. From a trader's point of view, the focal point of the Californian coast must be the Bay of San Francisco. This large body of water is one of the world's great natural harbours. Access from the sea is possible only through a narrow neck of water. When gold became the region's major industry, this entrance was to become known as the Golden Gate. The bay receives the waters of two large rivers, the Sacramento and the San Joaquin. These two waterways flow through the fertile plain of California's Central Valley. The Sacramento runs from north to south, the San Joaquin from south to north. On the way they both pick up the smaller rivers that flow down the western slopes of the Sierra Nevada mountain range. This range sits astride the border between California and what is today the state of Nevada. Winters on its slopes are characterized by heavy snow as well as heavy rains. In 1848 the western side of the Sierra Nevada was a forbidding place. The forest was dense, the terrain rugged and huge boulders studded the steep slopes. No tracks existed through the tangled scrub. It was in and around the rivers on the western slopes of the Sierra Nevada that the gold of California was found.

When the Mexican War broke out, the US army occupied California. With their own military in place, a growing number of Americans resolved to settle in that distant territory. In 1846 a series of wagon trains carrying pioneer settlers and their children made the epic 2,000-mile journey across the prairie, from the Missouri River to the Californian coast. More followed. Not all of the parties continued right through to the coast. After the murder of their leader Elder Smith, the members of the Church of the Latter Day Saints moved away from the hostile environs of Illinois and established the Mormon settlement of Deseret in the area of the great salt lake of Utah.[13] Many of the California settlers gravitated towards the shabby but well-situated town of San Francisco (then known as Yerba Buena). The subsequent growth of population and commerce in that town created a demand for all sorts of goods. It is here that the name Johann Augustus Sutter enters the pages of history. Sutter, a merchant of Swiss descent, held a substantial landholding at the junction of the Sacramento and American rivers. The American River was one of the larger of those that flowed out of the Sierra Nevada. Sutter had as his home a fort – Fort Sutter – that was something of a landmark in the area. While his main occupation was the raising of cattle, he was an ambitious businessman prepared to engage in all manner of other ventures. When it came to his attention that the settlers in San Francisco needed lumber to build their houses, he decided to add a sawmill to his string of assets.[14]

The event that led to the discovery of Californian gold is economically described by John Sutter's diary entry for 21 July 1847: 'Marshall and Nerio left for the Mts. on American Fork to select the site for the sawmill.' The Marshall in the diary note is James Marshall, an employee of Sutter. Marshall did, it seems, a very creditable job of constructing the sawmill alongside the South Fork of the American River. It was finished in January the following year. The mill was initially hampered by insufficient water-power, and Marshall decided to deepen the channel that ran beneath the waterwheel. No doubt he improved the operation of his mill. His excavation was also to have a more far-reaching significance. In the deepened millrace of the sawmill, on 24 January 1848, Marshall found some flakes of gold. There was some confusion at first, as none of the party could be sure that the yellow specks were in fact gold. The uncertainty is captured in the journal of one of the mill workers: 'This day some kind of mettle was found in the tailrace that looks like goald.'[15] The story goes that Marshall pounded the flakes with a hammer, boiled them in lye and compared their colour to that of a gold coin. His thoroughness was such that by the time he set off for Fort Sutter to report to his employer, the uncertainty had been dispelled. Marshall knew he had found gold.

Sutter tried to keep it quiet. Gold, he knew, would upset the orderly way of life that he had created for himself and his community. He doubled the pay of his men and cautioned them to say nothing. The stratagem worked – for a couple of months. The San Francisco newspaper, the *Californian*, printed in

March a small article that among other things mentioned a gold discovery in the 'newly made raceway of the Saw Mill recently erected by Captain Sutter on the American Fork'.[16] Still, little happened until Sam Brannan, a storekeeper living at Fort Sutter, decided it was time to drum up some business. He made his way over the mountains to the mill and from there he returned with a bottle full of gold dust. This he paraded through the streets of San Francisco. It had the desired effect. The last entry in Sutter's diary was made on 25 May: 'Great hosts continue to the Mts.' We can only imagine the chaos that thereafter engulfed the lands of this methodical Swiss.

Gold fever had hit the sleepy coastal towns. The dirt streets of San Francisco were emptied and ships were abandoned in the harbour. On 27 May the *Californian Star* published its last issue with the words: 'the stores are closed and places of business vacated, a large number of houses tenantless ... [it is] as if a curse had arrested our onward course of enterprise'.[17] The paper's editor then did the only sensible thing – he packed his bags and headed for the diggings. He was joined two days later by the editor of the rival journal, the *Californian*. By July there were several thousand miners panning the American River and its tributaries. Many made their fortunes within a few months. By November the word of this bonanza had reached the Eastern states. At first it was not believed. Then James Polk, President of the US, reported in his December message to Congress: 'It was known that mines of precious metals existed to a considerable extent in California at the time of its acquisition. Recent discoveries render it probable that these mines are more extensive and valuable than was anticipated. The accounts of the abundance of gold in that territory are of such an extraordinary character as would scarcely command belief were it not corroborated by the authentic reports of officers in the public service'.[18]

New York's newspapers began shouting the stories of incredible riches just waiting to be plucked out of the stream beds. It was then that the real rush began. Boats set sail from the ports of the East full of goldstruck fortune seekers. The first gold rush vessel to enter San Francisco Bay was a steamship named *California*.[19] It had left the Eastern seaboard before the great news had arrived there, but by the time it reached the level of Panama on the Pacific coast it encountered a crowd of fortune seekers who had crossed the Isthmus of Panama and were looking for a boat to take them north. The captain was obviously not a man to spurn Lady Luck. He loaded up with as many as he could hold and proceeded on northwards. His passengers were the first of an estimated 90,000 who arrived during that wild, optimistic year.

THE FORTY-NINERS

Getting there was the first challenge. About half of all the arrivals in the first year made their first sighting of San Francisco as they sailed through the Golden Gate. That is, they arrived by ship. The safer sea route was to sail around Cape Horn. Instead of stopping at Valparaiso, as the Chile-bound trading ships had long done, the packets and clippers simply continued to the newer port of San Francisco. The Cape route had one drawback. It took too long. Many others, less patient, took a shorter route. They sailed to Panama and, as the *conquistadores* had done centuries before them, disembarked and on foot or on mule they crossed the disease-ridden Isthmus. They expected, and indeed they had been told, that once on the other side it would be an easy matter to secure a berth on a boat from there to San Francisco. They were disappointed. Arriving on Panama's Pacific shore they encountered throngs of likeminded adventurers. Taking their place in the queue they had little choice but to sit and wait or try to bribe their way onto a northbound vessel. Death through malaria and cholera was common in the rain-drenched, crowded and mosquito-ridden camps.

There were simply not enough boats to go around. The Pacific coast of North America needed more than just a few months to transform itself from an irrelevant backwater to a vital and bustling trade route. Yet by year's end, the harbour of San Francisco was one of the busiest in the country. On landing in San Francisco, the would-be miners found a town fast becoming a city. The streets and stores were crowded and the rate of construction frantic. The original 200 wooden shacks and mud-brick huts could not long suffice for a population that by the end of 1849 was home to over 15,000 people on any night.[20] The town was also virtually the only port of entry for the goods needed to supply the miners up in the mountains. Then there was the gold that had to be sent the other way. Businesses of all kinds sprang out of nowhere. Despite the temptations and their long journey, most of the new arrivals stayed long enough only to get their bearings, stock-up on whatever provisions they lacked and secure a berth on one of the river steamers headed for the equally fast-growing tent city of Sacramento.

For each man who arrived by ship through the Golden Gate another came overland, making his way by wagon train across the plains. Throughout the first six months of the year a continuous stream of wagons converged on the banks of the Missouri, gathering in the towns of Council Bluffs, St Joseph and Independence, where they rested in preparation for the long trek west. While some veered to the south and followed the Santa Fe Trail through to southern California, the majority went west following the Oregon Trail through Nebraska, Wyoming and down through the Mormon lands in what is now Utah to central California. The early starters set out on the trail in the spring. The tardiest were just crossing the Missouri as the middle of summer approached. While there is no official record of the number of wagons that set out for California, it was at least 6,000, and possibly as many as 10,000. Most of those who set out were totally inexperienced. They learned as they went along. The going was never easy. At the outset cholera dogged the trail, claiming the lives of thousands.[21] Wagons were bogged, wheels broke and oxen died. The way became strewn with discarded possessions as each team strove to lighten the weight of their loads. And this was only on the first leg of the journey.

Situated on the border of Wyoming and South Dakota, Fort Laramie marked the end of the prairie. From here there were still well over 1,000 miles of mountain and desert to overcome. Beyond Fort Laramie the cholera inexplicably disappeared, but the travelling got even harder. Each day was a battle through mountain passes far more suited to mules than ox-drawn wagons. There was a brief respite when they descended the western slopes of the Rockies and entered Mormon territory. Then came the desolate and heart-breaking slog through the semi-desert. The constant need to find food and water was magnified for those who started late. When the desert finally came to an end, one terrible test remained. This was the climb over the northern foothills of the Sierra Nevada and the descent into California. The way was impossibly rugged. Many a wagon had to be abandoned and the journey resumed on foot. To add to the problems of those further back in the train, there was the fear of the winter snows. Everyone knew of the fate of the Donner party. This pioneer group had started the mountain crossing too late in the autumn of 1846 and found itself snow-bound. Of the 80 in the party only half survived to the spring. Many had done so only through eating the corpses of their dead comrades. In 1849 the snows were fast approaching before many had reached even the eastern slopes. The potential was there for even greater tragedy than had befallen the Donner party. But through luck and the timely intervention of the California militia, none such occurred. The great Forty-niner wagon train, battered and weary, got through substantially unscathed.

That summer and autumn an almost unbroken line of wagons had passed across a vast land that hitherto was unknown to Europeans apart from trappers, soldiers and hardy pioneers. An observer who remained stationed at any particular point along the trail would have seen perhaps 60 wagons pass by every day for a period of four months. The gold seekers, putting it in modern terms, had driven a six-lane highway from one side of the continent to the other. Never mind that the lands through which they passed were supposed to be Indian Territory, off limits to settlers. After the passage of the Forty-niners, it was only a matter of time before the American West was well and truly a part of the United States.

Most of the Forty-niners shared the nagging worry that by the time they arrived at the goldfields all the choice spots would be taken. At first the digging was concentrated in two rivers, the American and the Feather, and their tributaries. There had been plentiful gold and everyone could find a well-situated plot.

When in June of the following year the Forty-niners began arriving, it became plain that the two rivers could not hold such numbers. The new arrivals were forced to fan out all along the Sacramento river system. New deposits were found, and some far outshone the early strikes. Indeed, one of the richest and most extensive finds of the entire Sierra Nevada field was found by some displaced miners who were prospecting along Deer Creek, a tributary of the Yuba River. Finds such as this meant that the American and the Feather rivers were joined by the Yuba and Bear rivers as the centres of activity. Other prospectors pushed south along the rivers that drained into the San Joaquin. Though they were the scene of some enthusiastic rushes, these 'southern mines' never rivalled their northern neighbours either for gold or population. By the end of the year a 120-mile stretch of the Sierra Nevada was dotted with a myriad of camps large and small.

There was precious little government apparatus in place to help organize and somehow to manage the influx of this mass of humanity into the remote vastnesses. The only formal authority was a small military occupation force. As the land belonged to the US government, the President and Congress gave consideration to governmental regulation of the fields and the imposition of a system of licensing. After due debate the collective opinion was against it. It was no doubt a wise decision. The general in command of the Pacific Division of the US Army summed it up thus:

> I do not conceive that it would be desirable to have the mines worked for the benefit of the public treasury. To do that would require an army of officers and inferior agents, all with high salaries, and with opportunities and temptations for corruption too strong for ordinary human nature. The whole population would be put in opposition to the government array and violent collisions would lead even to bloodshed.[22]

THE GOLDFIELDS' DEMOCRACY

Thus the miners were left alone. Each camp made its own rules about how digging and other necessary aspects of life should be organized. The first priority of a group as they settled along some stretch of riverbank deep in rough forest was to create a code governing how claims could be staked. They were no experts in the matter. Most of them were relatively young men – between the ages of 18 and 35 – who had grown up in the cities or on Mid-West farms. Others came from France, Germany, Russia, Australia, Mexico, Chile and Britain and elsewhere besides. Some of the new arrivals were experienced miners, but most of them had only their wits on which to rely. They copied where they could and over time a code of considerable uniformity was developed. The code had as its basis two principal tenets: a man could claim only such ground as he could work himself, and he must continue to work it or risk it being claimed by another. The laws were designed, consciously or not, to keep out the capitalist. No absentee owner could lay claim to large plots and employ men to work them for his profit.

Each camp elected an official, the *alcalde*, whose task it was to arbitrate claims disputes.[23] Difficult cases could be settled by throwing the question to a general assembly. The simplicity and immediacy of this system was in part reliant on the fact that the economic life of the community was a very simple and uniform one. The business of nearly everyone was gold-digging. This frequent resort to general assembly has given the Californian diggings an undeserved reputation for discovering an idyllic form of 'pure' democratic government. Such an idealized view ignores, for one thing, the violence and racism shown by those of European descent to the Indians, Mexicans, Chileans and later the Chinese who joined the rush. Some mining codes banned non-whites from ownership of a claim. It was not for nothing that these ethnic groups formed their own camps, usually in the south, where the Europeans were less likely to persecute them.

In the early years of the rush, for the most part honesty prevailed on the fields. Gold was abundant enough for all. Certainly there was no shortage of drinking and gambling in the camps. The restraining influence of family life was almost totally absent. Yet men could leave their small stash in their tent or cabin while they worked their claims. There were, however, always those who would obtain gold by means other than the sweat of their brow. For such types the diggers devised their own ways of dispensing justice. Faced with a serious crime such as theft or murder, the cry would go out summoning men from all about until a good-sized group had been assembled. A trial would then begin at which the *alcalde* officiated. These were frequently perfunctory, the accusation sometimes being put to the assembly and a sentence passed by acclamation. The idea of granting the accused any formal legal representation was scorned. There was no right of appeal. There was little time or inclination to consider extenuating circumstances or to ponder the finer points of jurisprudence. If the accused was found guilty his sentence was usually brutal. The penalty for stealing gold was a whipping and expulsion. Occasionally the thief was branded or had his ears cut off to ensure that his conduct would be known to other communities. It was not unknown for him to be hanged. Indeed, in some camps theft was a hanging offence. Bidwell's Bar, at the junction of the Feather and Sacramento rivers, was one such camp, if the journal entry of Ross Clark, MD, can be relied on: 'Men seldom steal even a pick or shovel here, the penalty being death without delay ... When caught in the act, *up they go*, and that's the end of it.'[24] It was only with difficulty that the hangings and mutilations were suppressed when formally appointed sheriffs moved into the camps.

The mining codes provided a stability that the men themselves could not. The camps were far from being home to settled populations. In the years following the arrival of the Forty-niners, many found little enough gold. One problem was that the fields had become overcrowded. The work was hard and took its toll. Food was poor. Disillusionment set in quickly. As one expressed it in his journal: 'Two thirds of the people that are here would go home immediately if they had the means ... Generally speaking the gold fever cools down in a wonderful manner after a man has been here a week or two.'[25] The thing that kept them going was the hope of a fabulous find. The Deer Creek strike was one. Rich Bar, high in the mountains alongside a stream that emptied into the Feather River, was another. In the typically haphazard fashion of such discoveries, Rich Bar was found by three Germans who were trekking back to the

Feather River from a fruitless hunt further south. A rush to the area quickly followed. The claims at the Bar were limited to postage stamp size – ten feet square – in order to share better the wealth among the hordes who flocked there. The Rich Bar finds, however, were the exception. Mostly the lot of the digger was far less rewarding. By the end of the second year, fewer claims yielded gold in quantities that could secure a man's fortune. The jackpot was becoming harder and harder to strike. It was not that the gravels were becoming exhausted. On the contrary, gold there was aplenty. Rather, the depletion of the richest alluvials meant that to secure gold in large quantities a new form of mining was needed.

COMPANY MINING

The first to arrive at the American River came equipped with the prospector's pan and rocker. This latter device quickly gained in popularity. Its efficient operation typically required three or four men, but its benefit lay in the fact that the volume of gravel that could be washed was perhaps ten times as much as with a pan. The pan and the rocker remained the basic technologies throughout the first two years of the gold rush. When the richest alluvials began to give out, so began a steady evolution of the gold-washing process. The first step was to reduce the wastefulness of the rocker – a problem that was solved by placing mercury in its riffles. The next step was to divert the creeks and rivers so that the gravel in the middle of the stream could more readily be dug and washed. In this way many of the waterways became an obstacle course of ditches and walls. It was common to see a stretch of water divided into a series of parallel streams, any of which could be blocked off, pumped clear and the gravel within worked.

Increasingly, men looked for and found the gold-bearing gravel away from the water's edge. It lay in the 'dry diggings' up in the hills and in the ravines through which water ran only in the wet winter months. The dry miners had a choice – they could haul the gravel down to the water's edge or pile it up and wait for the rains. Either way, much of the poorer gravel lay unwashed, having been deemed not worth the effort. In its way the situation resembled the state of the South American silver mines before the coming of amalgamation. In California the solution was to prove far more simple. One of the richest and most extensive dry diggings was located at Nevada City, a tent community on the Yuba River. It was already the largest camp in California and it was to lead the way in the treatment of the dry gravels.[26] Nevada City's first contribution was the invention of the sluice. This was a wooden trough ranging from 20 to 200 yards in length. The sluice allowed the gravel from the dry diggings to be shovelled onto the head of the trough and then hosed down the boards in a swirling torrent that discharged its waste gravel into the river far below. The gold was recovered by a riffle arrangement at the bottom. With its introduction there was no longer a need to cart the heavy gravel down to the water's edge and then wash it laboriously through the rocker. The sluice was hampered by the fact, obvious enough, that there was frequently no water with which to operate it. Thus began the construction of overland channels through which water was conveyed from some distant stream to the dry diggings. The first such channel was one-and-a-half miles in length. A miner would pay to place a hose in the channel and draw off water to operate his sluice. It was the beginning of a new phase on the goldfields – for the first time men with capital to invest were needed.

Then, in 1853, a Frenchman working the Nevada City field experimented with placing a nozzle on the end of his hose and trained the jet directly onto the mountainside. The mountain melted away under the onslaught. Thus was born the practice of hydraulicking. It was the final refinement of the washing process. With this new technique, a single man armed with only a hose could send a good portion of a mountain down his sluices in only a few days. The practice had a devastating impact on the countryside. Whole valleys were turned into mud baths. But the gold was taken out and that was all that mattered. Hydraulicking had, also, a serious impact on the camps. Now one labourer could accomplish what only a few years before it would have taken 100 to achieve. Hydraulicking hastened the end of the gold rush days. After all, the operation of a rocker at the water's edge was a difficult exercise when mud was gushing down the slopes into the creek. Fortunately, hydraulicking needed more than just the one or two men required to wield the hoses. There were sluices to be constructed, aqueducts to be built and ground to be cleared. Gold was also found in the quartz rock, and soon the thud of black powder detonations and the crash of stamp mills could be heard through the Sierra Nevada. Companies were formed. By the middle of the decade the men working in these company operations outnumbered those who panned, rocked and prospected independently. By 1860 the small diggings were nearly a thing of the past.

Five years after James Marshall's discovery, where once there had been only a handful of sleepy villages, California was home to several large towns. Sacramento, standing on the site of Fort Sutter, was the distribution point for goods flowing into the northern mines. Stockton served the same role for the southern mines. One or two actual mining camps had also achieved considerable size – Nevada City was one, Yuba City another, the southern town of Mariposa yet another. Despite their luxuries and some elegant buildings, all these towns remained at heart mining camps. Tents were the majority of their dwellings. In contrast, San Francisco had become a business centre of growing national importance. The harbour was a forest of masts. The docks and warehouses were piled high with clothes, tools, canned food and building materials. Banks had sprung up to manage the affairs of the successful miners and the even more successful merchants. The postal service had become a major industry all of itself. Law firms proliferated as company mining gave rise to disputes over claims. And, increasingly, small manufacturing concerns were being established to serve not only the goldfields but also the needs of San Francisco itself.

This new economic order of towns and companies took time to reflect in goldfields society. The transition was a difficult one between the camps and the settled townships of a decade later. Crime and riots, in addition to the more traditional vices of gambling and heavy drinking, became a feature of the life as the miners came to terms with the progressive loss of their independence. Making matters worse, the camps filled with criminal elements from the East. Savage reactions eventually cleaned

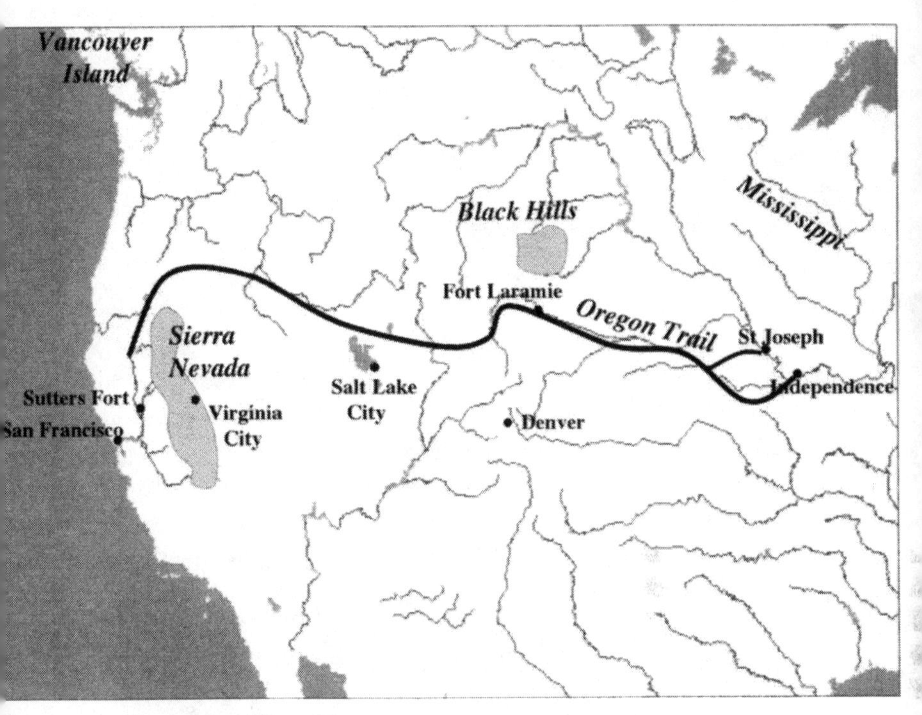

5 The American West, 1860

them up – in the meantime earning them a grim reputation for lynchings and mob violence – but only served to push the worst element into San Francisco. It took a couple of high-profile murders and the subsequent formation of the San Francisco Vigilance Committee to restore order to the city.[27] The next year the People's Party came to power in the city elections. It was to be a milestone in the establishment of capable, constitutional government in the state.

As the Californian economy matured, the business of gold extraction had increasingly to compete with other activities. The manufacture of mining equipment was one. Agriculture was another. The beginning of the end came when Edward Woodruff, a farmer in the Feather River valley, sued some of the hydraulicking companies for the flood damage caused by the silting-up of the Feather.[28] The case, and others like it, pitted the farm economy against the miners. After protracted and emotional litigation, a judgement was finally passed that prevented the discharge of the muddy torrents into waterways. Hydraulicking was finished. The hard rock mines lingered on, but they would eventually peter out. By the turn of the century the era of Californian gold was over.

EDWARD HARGRAVES

In the first years of the Gold Rush a total of about 3,000 fortune hunters had made their way across the Pacific to San Francisco from the isolated British colony of New South Wales.[29] Once in the Sierra Nevada, no doubt they met with the usual mixed fortunes. A good number would have failed. One of this latter group, a man named Edward Hargraves, was back in Australia by January 1851. He was not deterred by his ill-luck in California. Digging for gold was never going to be his road to success – he was too fat for that. His skills were that of a strategist and a promoter. No doubt he would have succeeded admirably as a merchant in California. Instead, while in California, Hargraves pondered the prospects of finding gold in Australia, at one point writing: 'I am very forcibly impressed that I have been in a gold region in New South Wales, within three hundred miles of Sydney'.[30] On his return he wasted little time in searching for that region. Hearing reports of several recent gold discoveries, and finding that these were consistent with his own recollections, Hargraves set out in February in the direction of Bathurst and the towns beyond. It took several months, but eventually, assisted by some local lads he had recruited, he found a few small grains of gold in a tributary of the Macquarie River. It was enough for him. Folding the few grains in a piece of paper, he headed for the Governor's mansion in Sydney to claim his place in history. The Governor was attentive, but no more. While in Sydney Hargraves received a letter from his local lads back west. This time they had found a true strike. It was the stroke of luck on which he had been relying. Returning west, he assembled a good collection of the gold nuggets and dust collected by his loyal team and travelled to Bathurst, where he spread the word of the great find. It had the desired effect. The men of Bathurst headed for the Yorky's

Corner, the site of the strike. The era of the Australian gold rushes had begun.

Edward Hargraves had imported the Californian gold rush into Australia. It is not so surprising that his role was necessary. It had needed Sam Brannan to parade his jar of gold through the streets of San Francisco before the Rush to the American River had occurred. And Brannan was living in American frontier society, famous for its individualist and entrepreneurial spirit. The Australian colonial society of the time was constructed along quite different lines. It had developed something of the class structure of contemporary English society. At the top was the Lieutenant-Governor, flanked by all the various appointees to senior posts in the colonial administration. Below these came the members of the city establishment – lawyers, bankers and wealthy merchants. The productive economy was firmly in the grip of the pastoralists. The south-east corner of Australia had become one giant sheep and cattle run. Much of the land had been settled without permission from, or payment to, the Crown. So common had 'squatting' become, as the practice was known, that it was almost respectable. The squatters were buoying the colony along with their sales of wool, and their prosperity was essential to the welfare of the economy. Economic dominance eventually brought with it political influence, and the colonial administration had come under the thrall of the pastoralists. Their main aims were to preserve their hold over the land and their access to the ear of the Governor.

It was a time when growing numbers of the common people of Europe were choosing emigration in preference to the hardships of home. The great majority went to North America. In most years Australia attracted no more than one in twenty of the emigrants who departed the United Kingdom.[31] One reason was its remoteness. Another was its popular perception as a gloomy convict-ridden pastoral land dominated by just the sort of landed aristocracy that many were seeking to escape. The Colonial Office continued to regard the Australian colonies as a dumping ground for convicts and a place of exile for undesirables.[32] Free immigrants, nearly all of them from the United Kingdom, who did make the journey had to compete with ex-convicts for what opportunities were allowed them by the entrenched colonial elite. American-style democratic government was unknown. Agents of social disruption were unwelcome in this society. Thus when gold was occasionally found the discoverer received little enough encouragement. 'Put it away Clarke or we will all have our throats cut',[33] is said to have been the Governor's response when shown a small nugget found by the prospector Pastor A. W. Clarke. The story may be apocryphal, but it has lived on by capturing the official attitude to the menace. When Thomas Chapman found gold in 1849, the men who flocked to his sheep run were dispersed by armed police. Few of the working class would have known how to negotiate their way through this official hostility. It needed someone who had seen how change could be wrought and who had the brashness and confidence to attempt it. Such was the role played by Edward Hargraves.

He was helped by the fact that the Australian colonies were ripe for change. Political divisions were opening up, driven by the desire of the town-based colonial businessmen to build a more broadly based, and therefore profitable, society. To achieve this they needed immigrants, and to attract them they had to be able to offer them land. The land, however, was held in the hands of the squatters, and this group had successfully resisted all attempts at a more equal distribution. Casting about for a different solution the reformers noted the copper boom in South Australia. They had seen how it had transformed and benefited the colony. An 1848 petition to conduct a mineral survey of New South Wales was sent to the Colonial Secretary, Earl Grey. He greeted it with caution. Some eighteen months later, perhaps influenced by events in California, Grey appointed Samuel Stutchbury as the Government Geologist for New South Wales.[34] So it was that when Hargraves roused the people of Bathurst with the thrilling prospect of gold, he could claim to have at least the tentative support of the colonial administration.

A GOLD COUNTRY

In May 1851 Australia had its first gold camp, named Ophir after the biblical gold town. Then in June a richer and more

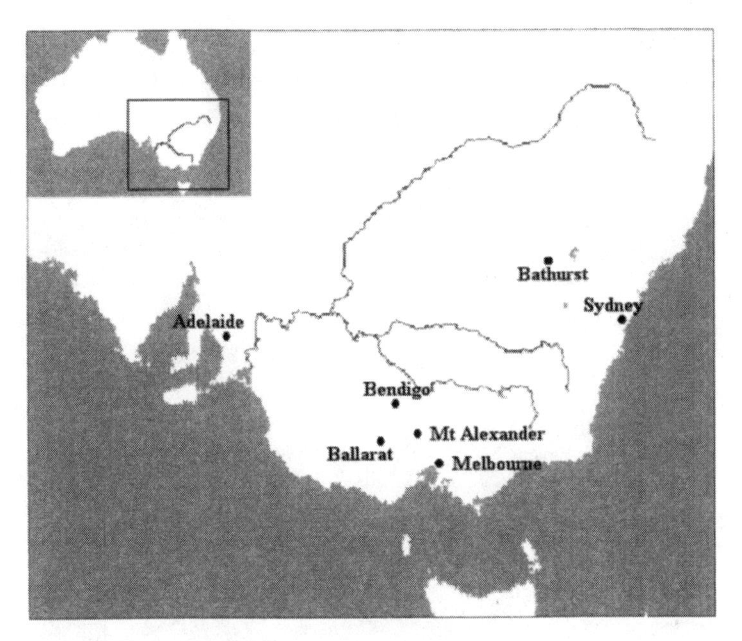

6 South-east Australia, 1855

extensive strike was made in the pastures along the Turon River 30 miles to the east. The next month a massive gold nugget found on a sheep run north of Ophir created a sensation throughout the colony. More diggers flocked to the area. Among them were several hundred from the new colony of Victoria. The exodus of their labourers was a serious worry for the leading men of the colonial capital, Melbourne. The town was already suffering from a business depression that had seen many depart for Sydney. In an effort to reverse the situation in June of that year Melbourne's leading businessmen gathered to offer a reward for the first payable gold strike made within 200 miles of the town.[35] It is said that the reward was claimed (unsuccessfully) before the meeting broke up. In any case, such an inducement was surely unnecessary. The hunt was on.

The first Victorian finds were unimpressive. By the end of July there were several hundred diggers working a sheep run near Clunes, a small town over 100 miles west of Melbourne. Weeks later gold in small quantities was found at Buninyong, a town lying on the road to Clunes. Next, gold was found at

Ballarat, on the road between Buninyong and Clunes. From the start this field was unusual, as little of its gold lay on the surface. Most of it lay in a thin seam between two and ten yards underground. Allotted tiny plots of eight feet square, the miners would dig straight down until they hit the seam. It was worth the effort. By mid-October there were over 5,000 diggers busy turning the pastures of Ballarat into a honeycomb. The burghers of Melbourne must at last have breathed easy. Victoria had its answer to the Turon River.

Ballarat was only the beginning. In October came a find in the valley below Mt Alexander to the north-east of Ballarat.[36] The rush that followed far outweighed any so far seen. By Christmas a tent city of 20,000 miners sprawled beneath the mountain. Ballarat was virtually abandoned as its miners scrambled to claim a plot in this new bonanza. Less than two months after the Mt Alexander rush came another rich find. This time the scene was Bendigo Creek, an out-of-the-way spot to the north of Mt Alexander. Gold, it was said, could be seen glistening in the creek bed. Half a dozen more significant finds were made in those heady months. And Victoria was not alone. In New South Wales gold was found in the mountains west and north-west of Sydney. By the end of the first gold rush year the two colonies could both lay claim to half a dozen or more thriving mining camps. The south-east of the continent, it must have seemed, was bursting with gold. Unlike in California, the rapid sequence of discoveries was due not to having a river system to guide the prospector's search. Instead, the diggers benefited from the fact that south-east Australia was already home to a large European presence. When news came of Ophir, these settlers were suddenly awakened to the possibility of gold. It lay almost literally at their feet. The sudden flowering of the gold camps brings to mind an image of a blindfold being removed from the eyes of the colonial population.

The rushes had a far greater impact on the smaller colonies than on New South Wales. While Sydney was abuzz with gold talk, relatively few of its male population actually decamped to the diggings. In Victoria, the smaller and less settled towns of Melbourne and Geelong were much harder hit: some quarters of the towns were emptied. South Australia, too, got the full dose of gold fever and lost up to half its men. Among them were virtually all the experienced workers from the copper fields. They were closed down for want of labour. Large numbers also set out from the island colony of Van Diemen's Land. The great majority of those who travelled from these colonies went to Victoria. There the strikes were richer and the journey – overland or by sea – was shorter. As it was less than a week's walk from the diggings, Melbourne was the point of entry for many of these fortune hunters, and soon that staid squatters' settlement had acquired the character of a boom town. In its taverns successful diggers squandered their fortunes, while its inns were host to thousands of hopefuls heading for the fields. To the more settled residents it seemed that the established order of civilized society was under threat. They had seen nothing yet.

The news took some time to reach the British Isles. A four-month delay between despatch of a letter from Melbourne and its receipt in London could easily be extended to six or seven if unfavourable weather hampered the ship in its passage. The early news that reached London of the gold finds had been sporadic and delayed, attracting little interest. To the mandarins in Whitehall a bout of gold fever in the antipodes was no doubt just another event, and a minor one, in an evergrowing list of colonial matters. All this changed when in April news reached London of the Mt Alexander find. Then the first ships carrying Australian gold sailed up the Thames. Overnight Australia became a destination with real attraction. Instead of heading for a new life in the United States, a large number of Britain's restless young men now sought a passage to Melbourne. The year 1852 was to the Australian colonies what 1849 had been to the west coast of North America. Where less than 19,000 had sailed for Australia in 1851, in the next year the figure was over four times that number.

Ships bound for Australia could not cope with the new demand, and a surge of activity overtook Britain's moribund shipyards. Seizing on the new market, American sea captains with their elegant clippers also took on the Australia run.[37] Once in Melbourne, for most men it was off to the diggings along the dirt roads leading through the mountains. When at last they arrived at Bendigo or Mt Alexander or one of the other fields, the new arrivals found that they were not alone. The diggings were filling up. The main fields were now tent cities, and the rich alluvials were giving way to poorer gravels. Most quickly realized that their fortunes were to be harder won than they had hoped. If this realization was shrugged off as they staked their claim, less easy to shrug off was the tax they had to pay for the privilege of being permitted to dig. By the end of the year resentment was building in the camps.

A LICENCE TO DIG

The colonial authorities had been caught off guard by Hargraves's bold stroke, and their first reaction was predictable: any activity that involved gold-digging must be very tightly controlled. Despite their support for a geological survey, the colonial elite still feared the disorder and economic disruption that gold rushes bring. Much of this fear had its roots in the upper-class distrust of the common man, particularly when he was confronted with so corrupting an influence as gold. The fact that many of the 'common men' were ex-convicts only heightened the apprehension of those whose task it was to maintain order and prosperity within that isolated colonial society. It is not then to be wondered at that when hundreds rushed to Ophir, the Lieutenant-Governor of New South Wales decided to impose on them a monthly licence fee.[38] The fee was, in itself, of little consequence. It was well justified in terms of legal precedent, and the diggers who flocked to the Australian goldfields never saw themselves, as did California's Forty-niners, as outside the reach of established government. A reasonable licence fee would have raised no objection. The trouble was that the licence fee imposed by Lieutenant-Governor Thomson at Ophir was never intended to be reasonable. Its size and the means of collecting it were designed to deter as many as possible of the fortune seekers. As a deterrent it was a failure, succeeding only in encouraging tax evasion and in irritating the great majority. Yet the diggings remained peaceful and orderly, and such was the apparent success of the fee that when gold was found in Clunes the Lieutenant-

Governor of Victoria, Charles La Trobe, adopted the same policy.

It was a miserable failure from the outset. The first year of Victoria's golden age was for its government one long humiliation. Policy reversals, digger protests, police brutality and corruption all became part of the scene. Such aggravation could perhaps be overlooked when winnings were plentiful. But as more people came to the fields and the yield of gold began to fall, there was less tolerance for what was seen as a rapacious tax underpinning a corrupt system. The situation came to a head at Bendigo when diggers banded together and refused to pay. Over the resistance of the squatter group – one of whom demanded that the diggers be tried for treason – the government decided on a large reduction in the licence fee and a promise to review the entire system. The news was greeted in Bendigo with celebration and a rousing three cheers for the Lieutenant-Governor. It was to be a shortlived support. When La Trobe was unable to capitalize on his brief popularity he was at last replaced. The incoming governor, Charles Hotham, was to fare no better.

Meanwhile, the mining scene in Victoria had begun to change. While Mt Alexander and Bendigo remained the premier fields, Ballarat returned to prominence. Here the gold lay in a network of ancient river beds hundreds of feet below the surface. To reach the rich gravels – known as leads – required perhaps six months of barren digging in sweltering and dangerous conditions. Even then the shaft might miss the lead and the men be left with no return for their efforts. The successful shafts, however, were often immensely rich. To carry out this type of mining a company of diggers was needed both to sink the shaft and to cut the wood necessary to line the shaft walls. Companies were also forming at Bendigo. In this case the challenge was not to sink a shaft, it was to liberate the gold from the heavy clay in which it was embedded. For this task was developed the puddling machine.[39] It was a simple enough device, consisting of a vat fitted with a long pole from which projected a series of paddles. A horse rotated the pole by circling the vat, thereby creating enough agitation to break the clay down into a sludge. In this way the gold was recovered. Puddling companies soon rivalled in strength the traditional prospectors.

Despite the evolution in the nature of mining, little changed in the administrative and legal machinery that governed it. La Trobe's reduction of the licence had created a breathing space, no more. Increasingly the real source of tension was the status of the miners and the legitimacy of their place in society. The more they felt themselves disrespected the more the diggers were willing to express their contempt for the government. The most common expression was fee evasion. The incoming Hotham was quick to grasp the situation and put in place a series of initiatives with the intention of recognizing gold mining as a legitimate industry, one that would no longer bear the flavour of a disreputable and somewhat subversive pursuit. Despite his insight his training led him into a fatal blunder. As a naval man of many years standing, Hotham could not tolerate the insubordination of fee evasion. He cracked down hard. Twice-weekly licence checks were ordered, and these soon became licence hunts. The temperature began to rise throughout the goldfields. In Bendigo there was renewed agitation, with reform being demanded in the three areas of the licence fee, the vote and the price of land.

The crisis came at Ballarat, following a flagrant case of judicial corruption. James Bentley, an ex-convict publican and a known bully, was acquitted of the murder of two drunken patrons. This led to widespread protests, in the course of which Bentley's hotel was burned to the ground.[40] Three men were arrested and convicted of arson, although hundreds had been involved. Mass protests followed in which political reform featured prominent among the demands. Hotham, convinced that he faced a political insurrection led by Chartists and revolutionaries, sent in the troops. (1848, the year of revolution in Europe, was still a vivid memory.) It was an explosive situation that was further inflamed by provocative behaviour on both sides. The diggers formed a ragged militia and dug into a stockade that had been built over the Eureka lead. Provoked and humiliated by the diggers, the troops stormed the stockade at four o'clock on a Sunday morning in December 1854. Twenty-four miners were killed in the fighting or in later indiscriminate shooting by the vengeful troops. The incident shocked the colony.

It was one of those events that bring to a confused situation an energizing clarity. Reform was swift and comprehensive. In March of the following year the licence fee was abolished. It was replaced by a gold export tax and a Miner's Right. By purchasing the Right, a man received not only the right to mine for one year but also the right to vote in the state elections. The administration of the goldfields was reformed. Locally elected officials were given more authority. The result was that the miners were effectively defused as a radical political force and became accepted into the mainstream of colonial economic and political life. As if to underline the fact, Peter Lalor, the leader of the Eureka militia, was elected as a goldfields representative in Victoria's Legislative Council, and later became a mine owner. In the 1870s he was recruiting Chinese as strikebreakers. In another way, too, the Eureka rebellion marked the end of an era. The wealthier miners began employing their less well-off colleagues in the activities of shaft sinking, wood-cutting and puddling. This trend towards company mining was reinforced with the building of quartz crushing plants at Bendigo and the sinking of ever-deeper leads at Ballarat.[41] By this time Melbourne had become a major city of the British Empire and Australia's banking capital. As for the gold that poured from its hinterland, the great bulk of it was shipped to London. It is reasonable to ask what purpose it served when it got there.

GOLD AND THE GLOBAL ECONOMY

During the 250 years after the discovery of Potosí, the world had known an adequate supply, perhaps even an abundance, of silver and gold. This state of affairs had been due almost entirely to the bullion of Latin America. The destruction that attended the wars of independence in the Spanish colonies saw a sharp decline in the production of those precious metals but there was no decline in demand. India and China continued, as always, to be insatiable in their appetite and the pull from these two populous regions meant that the bullion reserves of Europe drained away towards the Orient. They always had, of course, but now Europe's long-standing source of replenishment was failing and the continent began to experience a shortage of money. It came at a time when the steady growth in manufacturing production, agricultural output and other commercial activity was increasing the need for ready money. The result was that in Europe the general price level (which can be thought of as being the total money stock divided by the total of goods available) began to fall. Prices continued to fall until 1850 when the world was suddenly confronted with the flood of gold from California and Australia.[42] The price level revived. It so happens that the period of falling prices coincided with mounting social unrest in Europe, a turmoil that culminated in 1848 in a wave of revolutions that swept the continent. The decades that followed 1850, in contrast, were on the whole peacefully prosperous.

Was there any connection between the supply of gold and the general level of economic wellbeing? There were many contemporaries who thought so. In the preface to his *Critique of Political Economy*, Karl Marx remarked on 'the new stage of development which the [bourgeoisie] seemed to have entered with the discovery of gold in California and Australia'.[43] The 'new stage' to which Marx was referring was the growing political power enjoyed by the middle classes of Europe and the expansion of their industrial and commercial activity. To Marx, gold and the price recovery it brought had given renewed morale and strength to the merchant, the farmer and the shopkeeper.

What part did gold really play in this upswing in economic prosperity? There are some who, in agreement with Marx, attribute to gold a major role. Their case is based on the fact that the currencies of most European economies of the time, while paper-based, were 'convertible'. That is, paper money could only be printed if the issuing bank held a proportionate amount in precious metal. Thus the new gold allowed more money to be printed. This, say the supporters of this theory, had two impacts. First, it created a mild inflation in the United States, Great Britain and those other European nations whose currencies were backed by gold. After years of falling prices this gentle rise was a stimulus to business activity. Second, it enlarged the pool of funds that could be loaned for investment, thereby giving a start to budding industrialists and farmers wishing to improve their lands.[44]

Opponents of this theory contend that in reality gold was but one, and not a critical one at that, of the ingredients in the prosperity of the years 1850 to 1875. Events in Europe were moving in favour of economic growth regardless of the supply of gold. The yellow metal had, for example, little to do with the invention and spread of railways or of Bessemer's steel mills. Further, they dispute the claim that the revolutions that erupted across the continent in 1848 were a function of falling prices, contending that they were actually the outcome of the long and increasingly successful struggles of the bourgeoisie against the entrenched interests of the nobility and landed gentry. In these and other developments, the argument goes, gold was but a sideshow, of lasting benefit only in that it provided the catalyst for the settlement of new lands.

Debates on these matters are the preserve of specialists, and such issues, it may safely be assumed, were far from the minds of the majority of prospectors as they went about their daily work. For many of these, their attention was now directed at new horizons. The next stage of the gold rushes was beginning. Prospectors were beginning to spill out of California as early as 1855. They were the vanguard of a movement that would in twenty years see fortune-seekers push into every corner of the American West.[45] Many went north to Oregon. Others went south. News of a strike up in the Canadian colony of British Columbia led to thousands boarding steamers and heading for Vancouver Island, from where they made their way up the Fraser River to the diggings. It was a shortlived affair, distinguished only by the struggle for supremacy between the Californians and the local British colonial administration. In the following year gold was found in Colorado, in ravines located high in the Rockies. The gold had originally been reported by some Forty-niners who found traces of it near one of their camp sites as they made their way from Kansas to California. A decade later, 25,000 prospectors were working the Colorado fields of Gregory Gulch and Clear Creek. Just south of the Canadian border, other prospectors were making their way west into an unknown interior. There they found the gold that lay in the streams of the inland river systems. First came the Boise camp in Idaho, then some itinerants on their way to Boise from Colorado stumbled on alluvial gold in Montana. They named the site Bannack City. Shortly after, a far richer deposit was found at nearby Alder Gulch. Virginia City grew up close to the site, and soon an estimated 12,000 were working the Montana fields.

In the midst of it all, a generation of bitter dispute over the slavery question finally erupted into Civil War. The West was solidly behind Abraham Lincoln. Union army units joined the prospectors. They were needed – not to fight the Confederates, but to conquer the Indian tribes who were beginning to react forcefully to the brutality and arrogance with which their lands were being stolen. President Andrew Jackson had designated as Indian Territory all the lands west of the Mississippi. The Act was violated even before it could be enacted, as settlers swept into Iowa and Wisconsin. Gold in California provided the next large incursion into Indian lands. In 1851 a belated treaty was signed at Fort Laramie, allowing whites to establish roads across the region. It was when the Californians began to move inland that the violence began in earnest. A war broke out in Washington Territory, ending a few years later in total defeat for the Nez Perce. It was to become a familiar story. The Cheyenne and Arapahoe tribes in Colorado were pushed to desperate measures by the miners who streamed into their lands. They had tried initially to compromise. An Arapahoe chief even visited the small town of Denver, where he learned how to eat with a knife and fork. Such efforts were wasted. Wherever gold could be found, the miners poured in. The army backed them up, allowing their hatred and fear of the Indians to justify all sorts of treachery. Perhaps the most notorious incident of the era occurred on a November morning in 1864. As dawn broke that day, a heavily armed troop of 700 soldiers and militia descended on the Sand Creek campsite of a group of Cheyenne. Their leader had been promised safe passage for his people. He had been tricked, and by midday about 150 people, mostly women and children, lay dead. Their killers, happy with their work, returned to Denver to a rousing welcome.[46]

The intruders did not have it all their way. Few prospectors mustered the courage to stray far from the established camps. In the end, however, the guns and aggression of the whites were too much. Tribe after tribe was herded into reservations when the massacres had weakened them to the point of surrender. The Sioux were among the last to fall. Their final demise was put in train when gold was found by trespassing prospectors in the Black Hills of Dakota. This beautiful mountain region was the sacred land of the Sioux, protected by treaty with the US government. The army could not hold back the prospectors, and soon they were swarming across the Black Hills. Soon a multitude of miners were panning the streams of the area. Negotiations were tried to secure their safety. When they failed, as they inevitably would, the army sent in General George Custer at the head of a mounted battalion. The Sioux saw it as a declaration of war. In September 1876, in the valley of Little Bighorn, Custer and his men were surrounded and massacred. The aftermath was predictable enough. The independence of the Sioux did not long survive their victory.

In Australia, too, a second wave of gold rushes swept the eastern colonies. In contrast to the American prospectors, the Australian miners followed in the wake of the sheep and cattle. In the nearby colony of New Zealand a great expansion of sheep farming had blanketed the island. It is no coincidence that soon after gold was found on the south island. Veteran diggers crossed the Tasman Sea in their thousands, and for a brief time the New Zealand fields held a place of global importance. Sheep runs also spread far into the Australian colony of Queensland. Discoveries followed. First came Gympie just to the north of Brisbane.[47] This was soon followed by finds at Ravenswood and Charters Towers. In this way, gold seekers spread throughout the American West and Australasian colonies. It comes as some surprise, then, to learn that the greatest mining camp of the gold rush era was not a gold mine at all. It was a mine of silver.

THE COMSTOCK LODE

While the full force of the rush raged in California, on the eastern slopes of the Sierra Nevada a small isolated gold camp came to life. It nestled in a ravine the locals named Gold Canyon. There it lay forgotten by all but its 200 or so diggers and the Mormon traders who supplied them. The arrival, in 1858, of a few experienced men from California saw the beginning of mining in the quartz outcrops of the area. These workings soon failed, and the small field looked like it had reached the end of its life. It was fortunate then that a new strike further to the north was made at the end of that year. The men of Gold Canyon, with admirable practicality, named their new find Gold Hill. It was a rich patch of ground, and those who early staked a claim were soon making their fortune. A pair of late-comers began digging on the side of a nearby mountain named Mt Davidson. It was poor ground. Or, at least, it seemed so. That was until one of them tried washing a rocker-full of the odd-looking black-speckled pale yellow sand that lay in the stream beds. It was worth a try. Once washed, the sand that was left sitting on the bottom of the rocker gleamed in the sunlight.

A day or two later, as the men would afterwards recall, into the midst of this happy scene rode Henry Comstock. History has blackened his name. He has been called lazy and a drunk, a man 'always on the watch to avail himself of the exertions of others with as little fatigue to himself as possible'.[48] This accusation can be put down to the heated and vicious litigation that came along years later when it was realized how rich in silver was that yellow sand and the rock beneath. Comstock did what thousands of others did more anonymously throughout the gold rush era. Somehow he managed to convince the incumbent prospectors that his was the prior claim. Such was the vagueness of mining custom in those days that who could challenge him? The men formed a partnership. For the next month the sandy ledges of the canyon were dug and washed. As the little band dug deeper, it encountered an unusual black rock. The men worked around it. Then a visitor to the diggings had the idea of taking a sample to Placerville, back over the hills in California. There he had it tested. The results revealed a silver-bearing mineral of incredible richness. Henry Comstock and his partners had struck what history has come to call the Comstock lode.

Word soon spread in Placerville of this wondrous find from

across the mountains. By this time the Californian fields had come under the sway of shrewd and tough speculators. When the news of the strike made its way back to Gold Hill it was accompanied by some of these men. They convinced Comstock and his partners to sell their stakes for what the latter thought an excellent bargain. With more experienced and capable businessmen now leading the way, equipment for grinding the black rock was hauled in by mules, Virginia City was founded and silver bars in their hundreds began arriving in Sacramento. By Christmas 1859 Mt Davidson had become home to a thriving community. In the following spring it became clear that word had spread far indeed. A crowd rushed across the still snow-bound mountains from San Francisco and the inland towns. In a matter of months Virginia City became the boom town of the West. This of itself meant little enough. Boom towns came and went. The successful camps are remembered, while forgotten are those that suddenly flared brightly and just as quickly died. The Comstock lode would flare with a blinding glare for twenty years before it burnt out, leaving Virginia City by 1900 a ghost town. In the process it lit up a whole new landscape of industrial mining. It was a landscape where mining was a matter not for the small man, or even for the wealthy businessman, but for the ruthless millionaire prepared to gamble everything and destroy all who stood between him and the immense riches of the minerals beneath the earth. The Comstock lode revealed to the world the great drama of the coming period. It was the pioneer field of modern mining.

Fifteen separate claims soon occupied two miles of the mountain ledge on which Henry Comstock had mined with his partners. They were each large and well-capitalized concerns funded and directed from San Francisco. Shafts were sunk while water-driven hammer mills and wood-fired smelters were erected on the banks of the Carson River that ran through the valley far below. The road to riches meant sinking a shaft into the black rock. But where was the rock? One observer described its disposition as like 'raisins in a pudding'. This vague knowledge of the geology posed problems not only on the mining side of things, it also opened the gate for intense litigation as the Comstock companies sought to fend off encroachments from latecomers. It is said that during a five-year period more than one part in ten of the proceeds from the Comstock was invested in court battles. To see why this was so, it is necessary to understand some of the fundamentals of the mining law of the time.

Governments both State and Federal had resisted any attempt to impose laws regulating mining. The law of the camp was recognized as valid, and the role of the court in any dispute was to adjudicate on the basis of this local law. In 1866 an Act of Congress had stated that 'The mineral lands of the public domain ... are hereby declared to be free and open to exploration by all ... subject to such regulations as may be prescribed by law and subject also to the local customs and rules of the miners'.[49] In other words, Congress had seen fit to enshrine in legislation the spirit of the Gold Rush democracies. When veins of gold were found in quartz outcrops up and down the Sierra Nevada, a new set of 'local customs' had perforce to be created in order to govern hard rock and underground mining. The new rules were based on the commonly observed phenomenon that in the quartz rock of California a vein of rich ore – called the lode – usually snaked along the surface in some identifiable line. The practice developed whereby each claimant was allowed to stake a claim covering a certain fixed distance along this line. Boundaries between claims lying along the lode were thus clearly established (providing the claims were properly registered).

Less clear was the boundary between the lode claim and those claims that may lie alongside it – the so-called lateral claims. The custom developed that the owner of any claim could pursue *laterally* any vein that lay within his claim.[50] Thus a mine started on one claim could legally extend into an adjacent claim, provided it was still following the same vein of ore. But if the vein of ore could be followed from one claim to another, which claim had precedence? To solve this problem there developed the Law of the Apex. Whichever claim held the highest point, or apex, of the vein possessed also the right to pursue that vein onto all other claims. Confused? Suffice to say that among the quartz veins of California this approach worked adequately well. When more complex ore bodies were found, the laws broke down. As a result, the mineral fields became a lawyer's paradise. Colossal and endless legal tussles became a common feature of Western mining until 1918, when a long overdue Act of Congress mercifully abolished the system.

So it was that Eastern lawyers struck their own bonanza in Virginia City. For the most part the litigation was between the companies who had secured a claim along the line of the Comstock lode and the so-called vampires who could not get a foothold on the main lode but instead took up the ground adjacent. The courts were clogged, and hundreds of lawsuits stalled as three overworked judges attempted to cope with the load. Fortunes were at stake, and it seemed that violence must erupt without more rapid legal decision. Yet somehow the courts muddled through, and the mine owners, tired of the useless expense, came to compromise.

For the proprietors on the Comstock, of course, there was more to life than briefing lawyers. As they went about the task of expanding their operations they found themselves having to cope with the growing difficulties that the mountain held in store. Here they would frequently rely on the experienced and toughened men who were emigrating in droves from the dying fields of Cornwall. The difficulties faced on Mt Davidson help to explain why the Comstock became the breeding ground for the first generation of American hardrock miners who were then to take their skills further abroad into the mines and mills of the West and from there to the rest of the world. The first challenge to be encountered was the nature of the rock itself. The black ore was a crumbly mineral, and in the early underground workings the threat of roof collapse was an ever-present hazard to life and limb. A mine manager from California, Philip Deidesheimer, was commissioned to find the answer.[51] He came up with the technique of square-set timbering, a system of interlocking wooden beams that created for the miners a wooden safety cage, sometimes several dozen yards high. Though a glutton for wood, the square-set method would form the basis for underground roof support until new techniques were developed on a sounder knowledge of rock strength and behaviour.

A second challenge was the extraordinary heat of the underground workings. It is because of the intense heat emanating from Mt Davidson's subterranean rock that the Comstock must surely rank as one of the most malignant underground mining environments in history. Even at relatively shallow depths the rock became warm to the touch. When depths of 100 or more yards were reached the workings became oppressively hot and vapours given off by the rocks threatened asphyxiation in some of the poorly ventilated tunnels. In the first decade this was a difficult, but at least manageable, problem. Large fans were set up at the mouths of specially built ventilation shafts. In the second decade, conditions underground were to become hellish. Temperatures of 65 degrees Celsius were not unusual at the rockface. Men worked wearing nothing but a breechcloth, a helmet, packs of ice and thicksoled boots. Fifteen to twenty minutes digging rock was all they could bear before retreating to the merely simmering heat of a rest station.

The most pressing problem was neither rockfall nor heat, rather it was the oldest problem of all – the unstoppable flow of water into the workings. Mt Davidson, the miners were to discover as their shafts sank further into the earth, was riddled with subterranean streams. In 1863 the strike of a pick punched a hole through a clay wall in the Ophir mine and unleashed a spout of water that within two days created a subterranean lake seven yards in depth. It was a dramatic, but far from isolated, incidence of flooding. Pumps, some of them monsters for their time, were brought in from England. They failed to clear the workings. Production slowed. Some of the richer shafts were held up altogether. By 1865 Mt Davidson was threatened with a watery end – the same threat as had been faced by the Cornish mines of a century earlier. Who was to be the Comstock's James Watt?

Into the breach stepped Adolph Sutro with a plan simple and audacious. The essence of the plan was for a tunnel to be dug from the base of Mt Davidson to end directly under the Comstock lode. It would provide drainage for all of the mines and in addition would improve ventilation in the oppressive deep shafts. It would also allow for a cheaper method of delivering the ore to the smelters situated at the base of the mountain. It would be, of course, no small undertaking. The tunnel would have to be driven three miles through hard rock. It would need money and time. The mine owners jumped at the idea. Sutro and his plan were enthusiastically lauded by the Nevada State Legislature. The Bank of California, a shareholder in several of the mines, undertook to support the project if Sutro could win backing from the Eastern banks. Thus armed, Sutro set off for New York and Washington. And it is was then that his plan began to unravel.

ADOLPH SUTRO AND WILLIAM SHARON

Until recent times it has typically been the case that during the first years of a mineral field a crowd of independent operators would set up, labouring cheek by jowl along the lode. The provisions of mining law have ensured this. The riches of the earth, custom decreed, belong to everyone, and none should have more than their fair share. During the first years of the gold rushes this principle was strictly adhered to. Claims were small and one to a man. This changed when the nature of hard-rock mining demanded larger claims, but the size of the allowable claim was always much smaller than the full extent of the field. Obvious and gross inefficiency has frequently been the result of this profusion of companies. From time to time a visionary has stepped forward with a plan to cut through the mess and rid a field of inefficiencies and bring benefit to all. Sutro's Tunnel was one such vision. Where an individual with real vision has in addition a will of iron plus a formidable persuasiveness and ruthlessness to match, he is likely to succeed. Adolph Sutro, a mill owner on the Carson River, was one such man. Yet he failed, or at least his success came only when it was too late to matter.

Sutro was a Prussian immigrant who had arrived in the United States with his family in 1848. At the age of 21 he sailed into San Francisco, where for six years he ran a reasonably successful cigar and tobacco store. When news of the silver of Comstock rippled across California, he sold the store and left for the mountains. On the Carson River he built a mill with the intention of processing the waste from the larger mills. It was a telling move – a characteristic recognition that the practices of the majority of mill-owners were inefficient and remained profitable only because of the inherent richness of the ore body on which they lived. From his vantage-point on the fringe of the Comstock scene Sutro was able to survey the activity on the lode and form his conclusions with objectivity. He wrote to San Francisco's *Alta* newspaper:

> The working of the mines is done without any system as yet. Most of the companies commence without an eye to future success. Instead of running a tunnel from low down on the hill, and then sinking a shaft to meet it, which at once insures drainage, ventilation, and facilitates the work by going upwards, the claims are mostly entered from above and large openings made, which requires considerable timbering, and exposes the mine to all sorts of difficulty.[52]

In Sutro's mind there could be no excuse for working without an eye to the future success and, for him, his Tunnel was the pathway to that success for the Comstock. What could be more obvious? Urged on by all who heard him speak, he was on the verge of securing for his beloved project the backing of Eastern banks when a newcomer arrived in Virginia City. His name was William Sharon.

Sharon started his career as a merchant in San Francisco, lost his money in speculative investments and was hired by the Bank of California. His cool and capable style impressed the Bank's senior officers, and when the opportunity arose he was chosen as their man in Virginia City. It was an inspired choice. Sharon is not a sympathetic figure. Historians enamoured of Sutro's visionary scheme tend to place him in the position of anti-hero. For there is no doubt that Sharon blocked the Prussian's plans. He made multiple fortunes for himself and his employer while others were driven out of business by his actions. A record such as this does not make for popularity. Yet nobody did more to shape and strengthen the Comstock mines than he. The Bank was a latecomer to the Comstock, and when Sharon arrived in 1864 he found a town at the height of its glory. Two years later it was in the doldrums. Flooding had slowed the production from many mines, the plentiful black rock was becoming scarcer and the cluster of mills along the Carson River was half-starved of ore. Sharon began to lend aggressively to the struggling mill owners, undercutting the existing banks and accepting the mills themselves as security against his loans. As the lean times dragged on, one by one these mills fell into his hands, their demise hastened by Sharon's policy of rigidly refusing extensions of credit. By 1867 the Bank of California's branch in Virginia City was the owner of seven silver mills. But without ore they were useless, and in any case there were too many mills along the Carson River even for the good times. Sharon, a noted and successful poker player, may fleetingly have considered folding his hand. Instead he played his cards with such skill that within two years he was master of the lode.

The steps in which he moved forward form a progression with a compelling logic. First, the Bank's mills were grouped together into the Union Mill and Mining Company and shares purchased in many of the ailing mines of the lode. The ore from these mines was then channelled only to the Union mills, thus driving the independent mills out of business or into the arms of the Company.[53] The less efficient Union mills were then shut, while those larger or better designed were run at full capacity. Costs fell and output, once confidence was restored, rose to its previous heights. The Union Mill and Mining Company now virtually controlled the milling of Comstock ore. With such control now in his hands, Sharon began looking for further savings.

A railroad – the Union Railroad – was built from mountain mine to valley mill, eliminating the cumbersome carting by mules of the ores down the rugged slopes of Mt Davidson. And fuel for the smelters? A careful search for coal had been fruitless, and the slopes nearby were now completely bare of wood. Wooden aqueducts were constructed along which rushed a great torrent of water conveying trimmed logs from the more distant slopes of the Sierra Nevada to the banks of the Carson River.[54] An iron pipeline conveyed fresh water from Sierran streams. Cornish pumps sat at the head of the numerous shafts, as did the steam-driven hoists that daily lowered men down into the depths and at shift end raised them back to the fresh air. Their primary use was, of course, the hoisting of the ore. This whole industrial complex came to be known as Ralston's Ring, after its owner, William Ralston, the flamboyant chairman of the Bank of California. Virginia City took on the air of a company town. The near-monopoly of the Ring might have exercised its influence on the common labourers. They came together to form an industrial union in the summer of 1869.

The trigger for the movement was the fear of Chinese labour. The Chinese had become a feature of the gold camps in California and Victoria. Wherever he appeared the 'coolie' succeeded in arousing the hostility of the white diggers. Partly this was due to simple racism. There was little enough tolerance for non-European cultures at the best of times, and none was more different than the Chinese. They kept to themselves, worked relentlessly by day and at night indulged their passion for gambling and settling scores among rival groups. The hostility was also partly due to the knowledge that the Chinese were willing to work harder, and be paid less for it, than nearly any white man. Very often they would end up working the white men's discards. As a result of all this, anti-Chinese riots occasionally broke out. Stealing from a Chinese was not considered a particularly reprehensible act. They were considered fair game for the criminal element. It is easy to imagine the disquiet of the Comstock miners, then, when they looked down on the railroad construction from their vantage-point high up on the mountain and saw gangs of coolies laying track and clearing the way. The two declared aims of the union were to 'maintain our wages at a satisfactory standard, and to prevent the firm seating of Chinese labour in our midst'.[55] The latter aim was achieved when Sharon promised that once the railroad was complete the Chinese would be dispersed.

More surprising was how the union achieved the first aim. The principle that was eventually established held that all underground workers, no matter what their task, were to be paid $4 per hour. Skilled Cornishmen were to be paid the same rate as novices fresh from a Mid-West farm. And all underground workers must be members of the union. What more telling sign could there have been of the change in Western mining since the first free diggings on the American River less than two decades before? The union was tolerated by Sharon. In truth, he probably had little choice. Even in the days before the union no mine manager had dared to employ Chinese workers. Fear of worker retaliation was real. What Sharon could not tolerate, however, was an undertaking of the magnitude of Sutro's Tunnel, in which he was not the controlling partner. Beneficial or not, the Tunnel had to go. The Bank of California withdrew its support for the project.

The year 1869 was a low point. Production was less than half the level of four years previously. The Comstock was being mined out, and without the efficiencies brought by Sharon's conglomerate it might even have been in terminal decline. It was time again for the lode to bestow its unpredictable favours. A new strike boosted the silver output to nearly double what it had been in 1869. The field revived. The owners of the new find stayed resolutely outside the Ring's control. Others began to take an interest in trying their luck, and a battle for control developed between Ralston's Ring and the new groups. One of these was made up of four Irishmen – Mackay, Fair, Flood and O'Brien. They focused their attention on the oldest shafts, now all exhausted, which lay along the one-mile stretch of ridge that had proven the richest part of the field. In between two of them lay a patch of ground owned by the Virginia Consolidated Mining Company. Its shafts lay submerged and its pumps idle. No activity disturbed the surface. This forgotten claim was purchased by the Irish quartet, and down they dug. So contemptuous was William Sharon of their prospects that he even allowed them free use of the shafts in one of his own properties. He was not to know that the Virginian Consolidated held the biggest raisin of the entire lode. So rich did it prove to be that the four became the Bonanza Kings, the immensely wealthy proprietors of a mine that in its time would attract dignitaries and even royalty, all drawn by the allure of its fabulous wealth.

Was it a lucky strike? Or did the four harbour some knowledge of geology that was inaccessible to the great majority of speculators on the lode? The best interpretation is that it was an educated guess made by men who, unlike so many others, were prepared to look objectively at the field instead of being borne along by the various fashions, enthusiasms and panics. Precious metal strikes can be exaggerated, but the Virginia Consolidated is one instance when the description 'Aladdin's Cave' seems justified. As digging progressed, a great cavern, 100 yards from floor to apex, was excavated.[56] Visitors reported how the massive dome, supported by a forest of square-set timber, glittered in the candlelight. The Virginia Consolidated launched the Comstock on a career that made it unquestionably the richest mining field on earth between 1873 and 1878. The silver of Nevada flowed in such quantities that it helped propel the restructuring of the world's currencies.[57] Then, just as suddenly, it was finished. As a footnote on the value of perseverance, that same year Sutro's Tunnel finally punched through to the bottom of the Yellow Jacket shaft. Alas, it was too late to serve as anything more useful than as a morality tale. The decline of the lode was rapid. In 1881 it was producing no more silver than it had back in 1860. Never mind. For the American West, the Comstock had leaped the gulf between simple alluvial diggings and the world of large-scale and highly capitalized underground mining. It was ready to hand the baton to a new generation.

THE END OF THE FRONTIER

In 1890 the frontier was declared officially closed, although its every corner was known to Europeans well before that. The taming of the West was an incremental process. By the late 1850s, a network of transport links had come into being between the Mississippi and the California coast. Bullock wagons, mules, the Pony Express and horse-drawn carriages were used for the overland transport of people, mail and lighter goods. Bulk goods and heavy manufactures had to go by sea. Periodically a cry would be made demanding a railroad to tie together this great and growing nation cleaved into two by a sea of prairie. Lincoln ran for President promising a railroad to the Pacific Ocean. Each time the vision foundered on the rocks of parochial politics – through which towns would the railroad be built? – and on the reluctance to commit the funds for so large an undertaking.

Then came the Civil War. The mood in Washington swung towards nation-building, and for the first time a real consensus began to build in favour of a railroad linking the states east of the Mississippi to California. In the midst of the War, Congress passed the Act supporting this 'great military highway'.[58] The building of America's first transcontinental railroad took seven years. The plan was to start building inwards from both ends and meet somewhere in the middle. The Union Pacific started at Omaha on the banks of the Mississippi and the Central Pacific began at San Francisco. The Central Pacific will live on as an epic feat of construction. Thousands of Chinese labourers carved track ledges out of sheer cliffs, battered their way through mountains of solid rock and laid track in record time across the parched flat lands when at last they were through the Sierra Nevada. In May 1869 Leland Stanford drove in the golden spike that linked the two railroads. The United States of America had at last fulfilled what the journalist had called its 'manifest destiny'. It stood united from sea to sea.

As the great success of the Comstock brought confidence, capital and skills to the Western mining camps, the railroad, along with the telegraph, improved communications between the city and the remote camps and brought them into the orbit of urban financiers. The railroad also reduced the cost of supplies to the mines while providing cheap transport to markets for their produce. When the silver of Comstock had first started flowing in abundance, it had been devoted to bestowing on the city of San Francisco grand buildings, local industry and property development. Then the profits had begun flowing outwards into new mineral ventures. One of the bigger ones was in the Black Hills. Scarcely one year after the massacre at Little Bighorn, three San Francisco entrepreneurs – Haggin, Hearst and Tevis – established the Homestake Mining Company and built the nearby town of Lead City. Square-set timbering, Comstock-style, secured the underground workings, and rows of hammer mills were erected for breaking the quartz in preparation for amalgamation. Not long before, in nearby Deadwood Gulch, Wild Bill Hickok had been shot in the back and killed while playing a game of cards.[59] Modern industry had come to the lawless and remote fastnesses of South Dakota.

Sprawling cattle ranches were filling in the wide spaces between the gold camps and the dusty towns. There were others who wanted to fill in the wide spaces with still more mining camps. One of them was Clarence King, a Yale graduate who had specialized in the study of geology and mineralogy.[60] He had developed a fascination for the West, and with a friend spent time travelling through the prairies, working for a brief while at the Comstock and being chased by threatening but, it seems, none too deadly Indians. Like others of his time he could see the revolutionary possibilities inherent in the railroad. It could be the vehicle to bring the mineral potential of the West to reality. To assist this, he reasoned, just as the railroad had needed thorough surveys to establish the physical lie of the land, so the geological lie of the land must be mapped out by an equally thorough survey. With this vision in his mind and a plan in his pocket King set off for Washington. The legislative processes of the US Congress have always been open to influence by a determined individual with a special interest, and King spent a winter in the national capital convincing the lawmakers to pass a bill authorizing funds for a geological survey of the lands bordering the 40th parallel – that line of latitude passing through the Comstock, northern Utah, Wyoming and Colorado. He was only 25 years old when, funded by his country's government, he gathered about him a group of young engineers and geologists and despatched them through the remote and sparsely populated territories along the parallel. Their mission was to record the geological formations of the region and to document the mining activity spread across it.

In the end the survey would last ten years and its summary volume, published in 1877, occupied fully 800 pages. Its success inspired similar surveys. To the practical mining man the purely geological side of the survey, which could tell him such things as whether or not he was digging into Pre-Cambrian rock formations, was of little value. To the practitioner the far greater value of the survey was revealed in the episode of the Great Diamond Hoax.[61] This began with reports of a huge find of diamonds in a remote corner of southern Arizona. An office had been set up in San Francisco to promote the bonanza and sell shares in it to lucky early-bird investors. The diamonds were found in a location that had been explored by King and, intrigued by the discovery, he set off on horseback to examine the claim for himself. It took him little time to realize that the whole thing was a scam and that the sandy soil had merely been crudely salted. The exposure of the hoax created a sensation and made King a public figure, especially when it became known that the hoaxers had offered him a large inducement if only he would maintain a few days silence. The affair also alerted investors to the value of men whose scientifically based understanding of geological formations gave them special insight into what might truly be awaiting in the rock beneath their feet. It is no accident that throughout his post-survey career King was much sought after by antagonists in the Law of the Apex battles. His upright reputation carried a lot of weight with perplexed judges and juries struggling to make head or tail of the confusing barrage of claim and counter-claim, little of which could be in any way proven.

The West for long remained a wild place. In Arizona, the site of one of the many gold rushes of the time, the depredations of the miners and ranchers led to the last great Indian war when the Apache, urged on by their young leader Geronimo, broke out of their reservation.[62] Although defeat was inevitable, they kept up a running battle for two years with the US Army, thereby providing material for a legion of cinema Westerns. The violence was not confined to Indian skirmishes. Cowboys, mostly Mexicans, added another volatile element to frontier life. Robbery was common and gunfights a part of life. The shootout at the OK Corral has gone down in history, but similar events were occurring every week. It could not last forever. Slowly the West was absorbed into the mainstream of American life. The driving force was the country's burgeoning economy. It was a period of accelerated industrialization that built on the foundations of the infrastructure laid during the Civil War.[63] The United States had for decades shown the potential to be a world power. It was time for big business to make it a reality.

FOUR

The Copper Barons

BRIMSTONE

The province of Huelva lies on Spain's south Atlantic coast, pressed up against the border with Portugal. It is home to Palos, the Franciscan monastery in which Christopher Columbus spent his last night before setting sail west. Palos lies on the Rio Tinto. The river derives its name from the reddish tinge bestowed on its waters by the copper ores that lie near the banks along its upper reaches. We have already seen that the mines of this area were an important part of the metal supply of Imperial Rome. When the Vandals overran the region, ordered civilization collapsed and the mines were forgotten. The awakening of the Gothic age passed them by. It was only when the Spanish monarch Philip of Habsburg launched a search for domestic mineral deposits that they were rediscovered. Work began afresh in 1546.[1] It was a desultory affair, and an unusual one. Unusual in two senses. First, part of the copper was won from its ore not by smelting, but by dissolving it in tanks of acidic water. Once dissolved, the copper would be precipitated on iron rods placed into the tank. Second, that part of the ore that was smelted gave off great clouds of noxious fumes and smoke. The surrounding forests died away. The fumes were sulphur dioxide, known in modern times as the principal ingredient in acid rain. The ores of Rio Tinto were copper pyrites. Pyrites are essentially a combination of metallic elements and sulphur, with the latter in much the greatest proportion. Copper was all that interested those early smeltermen. Of what use could be the sulphur?

Enter Nicholas Leblanc. The seventeenth century was a golden age for chemical discoveries. The way was led by German and French scientists, graduates of the polytechnics that were beginning to offer scientific education instead of the time-honoured university curriculum of Latin, Rhetoric and other medieval fare. In keeping with the spirit of the times, the French Académie des Sciences offered a prize for the invention of a process to produce cheaper soda, an ingredient essential in the manufacture of soap, glass and bleaches. The prize was claimed by Nicholas Leblanc, who in 1791 succeeded in producing soda using just three essential ingredients.[2] The first was common salt, the second chalk and the last sulphuric acid. Leblanc never received his prize, nor did he even benefit from his process. Instead he lost control of it during the tumult of the French Revolution, when the Revolutionary government seized the rights to the process and decided to allow unfettered access to all comers. Perhaps they overlooked the fact that 'all comers' would inevitably include their arch-rivals, the British. It was they who benefited the most. Leblanc's process was adopted by Scottish industrialists, who constructed a series of large soda factories along the banks of Glasgow's River Clyde. Of the three ingredients, sulphur was much the most difficult to obtain. The only known large deposits of mineral sulphur were found on the Mediterranean island of Sicily. In this way the brimstone of Sicily grew in importance until it was supplying four-fifths of the world's sulphur.[3] Much of it was loaded onto ships bound for the foggy mouth of the River Clyde. On their return journey, the ships would bring North Atlantic herring to supply the tables of the devout southern European Catholics. It was a brisk trade.

A concentrated source of a vital industrial material has always been an open invitation to those who would seek to gain from 'cornering' the market. To such men the brimstone of Sicily must have presented a mouthwatering target. Speculators in Paris and London wrought havoc with the trade, buying up great quantities in order to force up the price and then releasing their hoards. After one particularly ruinous period of speculation, when the price was forced up threefold before plunging again, an exasperated King Ferdinand II of Naples moved to establish control. He first created a crown monopoly on the mining and sale of brimstone, then he increased the price. The response was swift and unambiguous. Lord Palmerston, the

British Foreign Secretary, sent a note to Ferdinand in which the threat was made clear. The monopoly, he stated, was an infringement of the rights of British subjects. While Ferdinand raged against such a high and mighty proclamation, the British seized his vessels. Neapolitan troops were shipped to Sicily. The stand-off continued until the King gave way and cancelled the monopoly. The price fell and the crisis was over. Why the fuss? Simply put, through his action Ferdinand had exposed one of the two great insecurities of industrial Britain – its dependence on imported raw materials to feed its factories. (The second was the fear of a factory workers' revolution.) The lesson was not lost on the British chemical firms, and the search was intensified for new sources of sulphur less vulnerable to manipulation. Soon it was realized that pyrites would provide the answer. The first to be worked for their sulphur content were those in Ireland's County Wicklow.

From there attention turned to the richer deposits of southern Spain. Easily the largest of them was at Tharsis, a mining complex that had been owned and operated by a French syndicate for 50 years. Sensing opportunity, the French seized their chance, aggressively lowering their prices until they succeeded in capturing much of this new market. Their profits were hampered because of the wastefulness of the pyrite-processing technique then in existence. Using the pyrites for soda-making was straightforward enough. Unfortunately, the copper remaining in the burnt pyritic ore had to be thrown away. So wasteful a practice was unlikely to be tolerated for long in industrial Britain. The solution was found by a Glasgow chemist, William Henderson.[4] He devised the process whereby the copper in the burnt pyrite 'cinders' was dissolved and then recovered by precipitation on iron. The first works based on the Henderson process were built near Newcastle upon Tyne. Soon similar plants were dotted throughout the major industrial centres of England. The process was so favourable to Spanish pyrites that soon a modern system of transport was needed to get them to port. The French owners, reluctant to commit the necessary capital, sold out to a syndicate led by the Scottish soapmaker Charles Tennant.

Tennant set about modernizing his acquisition. He constructed a railway from Tharsis to the port of Huelva and expanded the open cut mining operation. He purchased the rights to the Henderson process, and with William Henderson as a director, built several sulphuric acid plants in Britain. Tharsis became a formidable, integrated mining, chemical and metallurgical conglomerate occupying a position of dominance in the global sulphur market and ranking among the largest copper producers in the world. From his office in London, no doubt the thoughts of Charles Tennant occasionally strayed in the direction of the Rio Tinto mines now operated by the Spanish crown. Perhaps he pondered the possibilities inherent in their ownership. Or perhaps he felt relieved that they were so poorly managed that, despite their potential, they posed no threat to his own highly profitable business. For the pyrites of Rio Tinto were the real jewel of the Spanish pyrite belt. In June 1870 a cash-starved Spanish government finally took the step of offering them for sale.[5]

Spain had been weakened for over 50 years by civil strife between the state and the Carlists, conservatives who drew their strength from the rural areas. A turning-point in the struggle had been reached when Queen Isabella fled to France in the face of an army-inspired revolution. Into her place stepped the reformist generals Juan Serrano and Juan Prim y Prats. Prim had long been a guiding force in Spanish politics and it was he who assumed the running of the new government. His was the decision to offer Rio Tinto for sale – the reason being the eminently practical one that the government needed the money to help restore order. It was no light decision. The mines were a national treasure and part of Spain's heritage. To sign them over to foreign ownership was an admission of inferiority that was felt keenly in Madrid. But Prim was never a man to shy from hard decisions. He also took on the task of finding a replacement monarch for his country. It was another thankless task, one for which he was assassinated on the day the new German-born king entered Madrid.

It is understandable, in the light of the political situation, that the sale of the mines initially attracted little foreign interest. Making matters worse was the fact that the asking price was $4 million. And this was just to buy them. Another huge sum would be needed for railway and mine development. No other mine in history had demanded anything like so much. When the sale was advertised across Europe, it received no bid that reached the reserve price. The closest was from a consortium led by a Scottish businessman, Hugh Matheson. The 50-year-old Matheson was a respected figure in the City of London. His credentials went back to his involvement in the Hong Kong trading firm of Jardine-Matheson & Co. He had made investments in tea estates in India and railways and cotton mills in China. Now he was ready for a bigger challenge. He tried first by backing Baron de Reuter in his successful bid for the exclusive right to build and operate all railways and exploit all the mines and forests of Persia. While the Shah of Persia was later forced to withdraw the concession, such ambitious projects were henceforward the focus of Matheson's interest. Now he had found another in southern Spain.

Rio Tinto was unquestionably a treasure chest. The only real question was how much money could be made out of it, given its need for large-scale investment in railways and other facilities. It all depended on how one viewed the sales outlook for the project. It is here that Matheson's experience of the City financial community came into its own. He deftly built up interest and support amongst financiers large and small, undertaking an adroit public relations campaign and enlisting support for the project from newspapers and journals. His campaign had its questionable side. The project's prospectus claimed, in effect, that the growth in global demand for pyrites would be massive, and that Rio Tinto would supply most of it without seeing a reduction in price. Such claims amounted to deliberate deception, and there can be no question that Matheson and his partners had taken a huge risk with other people's money, all the while painting the rosiest of assured futures for the company.[6] In this way the Rio Tinto project set the mould for the market manipulations of later years. It also blazed a trail in a more positive direction, for Matheson's syndicate of bankers and railway contractors had succeeded in persuading a market long sceptical of mining ventures to pour huge sums into just such a project. In this, as much as in their questionable stock-market tactics, Matheson and friends became pathbreakers for a new era.

Their immediate obligation was the delivery of the sale price. They smuggled the initial instalment into Madrid via rail and cart, being careful to avoid the warring factions for whom the gold would have been a prize second to none. The railway and mine expansions were finished by 1875, and the first shipment left Huelva in the summer of that year. It was priced well below the going rate – a necessity, for Rio Tinto had first to break the stranglehold of Tharsis over the British sulphuric acid manufacturers. The ensuing struggle brought all the Spanish pyrites producers to near ruin, and they eventually came together to declare a truce and carve up the market between them. Matheson knew how to press his advantage. The Rio Tinto syndicate emerged from the truce with a good share of the British sales and with the European market as its own. The Spanish producers were not alone in seeing value in dividing up the market between themselves.

LAKES COPPER

A copper combination was also being formed in the remote north of Michigan. After the rush to Michigan's Keweenaw peninsular, the copper fever had petered out. The initial promise had proven difficult to convert into commercial success. One problem was the nature of the deposits. Instead of the usual vein of copper ore that could be followed into the earth, much of the copper of Michigan existed as lumps of almost pure metal. While some weighed hundreds of tonnes, their location was unpredictable, and many promising beginnings ended in barren shafts. Only two mines really prospered in the early years, the Minesota and the Cliff,[7] and even these remained modest because, in part, of a second problem. This was the prohibitive cost of transportation to the main smelting centres of Baltimore, Boston and New York. It was not that the Keweenaw was badly situated from a transport point of view. It was on the shores of Lake Superior, one of the largest inland bodies of water in the world. Transport through the Great Lakes system to the major cities should have been easy. One obstacle, however, stood in the way. This was the Sault Sainte Marie (St

Mary's Falls), the narrow rapids that separate Lake Superior from Lake Huron to the east. Shiploads of ore had to be offloaded from the vessels that plied Lake Superior, and then carried around the rapids before being loaded onto a waiting ship bound for the Eastern seaboard. The cost of transport was so high as to render the mines barely economic, despite their riches. The result was that, for years following the discoveries at Keweenaw, the Eastern smelters continued on their diet of Cuban and Chilean ores. Copper was not the only raw material to be so inconvenienced. Large deposits of iron ore had been found in the Marquette Range, a week's walking distance from the Keweenaw Peninsula. At that time much of the nation's iron and steel-making was done in the coal-rich states of Pennsylvania and Ohio, bordering Lake Erie. In 1852 a few barrels of Marquette iron ore were shipped to the forges there, but it was a small affair. As with Keweenaw copper, the transport costs were too high to justify any large-scale development. Yet clearly the potential was there, and momentum began to gather for the construction of a canal that would skirt the falls.

As with most matters of US public policy in those years, the subsequent development of Great Lakes mining now became mixed up with the dispute over slavery. A Sault Sainte Marie canal proposal made its way to the US Congress. There it met bitter opposition from the slave states, fearful as they were of any development that might strengthen the hand of the North. It was repeatedly blocked, but in end the public funds for the project were approved, and two years later the canal unlocked the mineral riches of Michigan's Lake Superior coast. Cornish miners facing a bleak future at home helped to introduce the modern machinery that assisted a rapid expansion. The outbreak of the Civil War provided another healthy boost. But when the Confederate General Robert E. Lee surrendered at the Appomattox Court House in the spring of 1865, the United States seemed to fall back exhausted. With the emergency of war no longer pressing the Union, and the assassinated Lincoln suddenly in his grave just days after Lee's surrender, an economic and spiritual depression gripped the country. Industrial activity plunged, forcing many Michigan mines to close or drastically to reduce their output. The region took on an air of abandonment. Yet a prospector's optimism can never be completely extinguished. Indeed, one historian has even put forward a theory that it is in years of economic depression that prospecting activity and hence mineral discovery is accelerated.[8] Edwin Hulbert was one such prospector. The metallic copper of Keweenaw was clustered at the tip and base of the peninsula, and, in consequence, the middle section had been left alone. In this heavily wooded wilderness, Hulbert uncovered a curious cement-like rock that was intricately intertwined with a stringy blanket of copper.[9] He called his find the Calumet. With backing from some Boston financiers, a young Harvard zoologist, Alexander Agassiz, took this and a neighbouring claim and developed them into the Calumet & Hecla. It would become the pride of the region.

It may at first seem odd that a Harvard academic specializing in life sciences should find himself charged with the responsibility of making a commercial proposition out of a difficult ore body located deep in a forbidding wilderness. There was a family connection. Agassiz's father, Louis, was one of the eminent geologists of his time. He had first come to public attention when he delivered an address to Switzerland's Helvetian Society. In the course of the address he set forward the theory that the earth had once, not so long ago, been in the grip of an ice age, during which massive glaciers had shaped and scoured the landscape of Europe. The audience reception of this proposition was, as one commentator remarked, as icy as the subject matter, and even the world-famous traveller and mining chronicler Alexander von Humboldt advised Agassiz to quietly drop the whole idea. The experience only strengthened his resolve, and for several years he gathered evidence from Alpine glaciers until he had collected sufficient to pen his classic work, *Studies on Glaciers*.[10] Rarely can a geological controversy have been resolved so rapidly and decisively. In later life he moved to Boston to pursue evidence of glacial action on the North American continent, and it was in that city that he formed the connection with financiers interested in the riches of the Keweenaw. Agassiz the younger possessed a similar resourcefulness and determination. Not only did he tackle the technical problems of the project, he proved to be a man of formidable political skills.

The United States has, in the popular imagination, acquired a reputation for being the champion of free trade. This reputation has rarely been deserved, and certainly was not during its industrializing years. During the post-Civil War depression, tariff protection was the populist policy of the day. The Michigan copper producers were among the more vociferous in clamouring for a trade wall to help them weather the storm. While they were struggling, shiploads of Chilean ore were being unloaded at New York and Baltimore. In response, the Michigan Legislature presented a petition to the United States Congress for higher protective tariffs on copper. The debate was skilfully conducted by a Congressman Driggs from Michigan. The tariff on iron and steel was hurting copper sales; therefore, he blustered, the copper producers needed one also. Neither was Driggs above using the argument – especially potent in the aftermath of the Civil War – that 'peons' (another term for slaves) were being used to produce the South American copper.[11] The Bill was passed by a huge majority, despite a Presidential veto. Within a year the measure had forced the Baltimore smelters, along with those in New York and Boston, to shut, and smelting moved west to the shores of the Great Lakes. Not content with so effective a step, the producers under the leadership of Calumet & Hecla organized themselves into a cartel. Thus protected by tariff walls and freed from domestic competition, Great Lakes copper entered a newly prosperous period.

FROM SILVER TO COPPER: INDUSTRIALIZING THE AMERICAN WEST

The career of George Hearst might well be considered a parable of the development of mining in the western United States. The son of a Missouri farmer, he joined the rush to the California goldfields, arriving there in 1850. During the next decade he made his living prospecting and running various stores, a typical enough career for a Californian miner. He seems to have missed the opportunities that arose when company mining was taking over from the individual prospector. Perhaps he learned from it, however. News of the Comstock bonanza drew him across the Sierra Nevada as one of the earlier speculators. He secured for himself a large stake in the Ophir mine, which he eventually sold at the peak of one of the lode's regular stock frenzies. Departing the Comstock, he tried his hand at several of the many other new mineral fields that sprang up as the Californian miners ventured inland in search of richer ground. Hearst lost much of his fortune during this decade, as one gamble after another proved to be barren. Then he struck up a business alliance with two San Francisco businessmen, James Ben Ali Haggin and his brother-in-law Lloyd Tevis. Haggin was a rarity amongst the business community on the West coast. A tough and shrewd financier, it was said that he had never failed in a venture. If true, this may have been due to the restraining influence of Tevis, who assumed the role of cautious banker in their partnership.

When Hearst joined Haggin and Tevis he seems to have brought an element lacking in the experience of the latter two. He had first-hand knowledge of, and excellent contacts within, the treacherous but potentially lucrative world of Western mining, and the syndicate was soon rewarded when Hearst was approached by his old acquaintance, Marcus Daly, an experienced miner with a reputation for 'being able to see further into the ground'[12] than any man in the West. The pair had first met at the Comstock, Hearst as mine speculator and Daly as mine superintendent. Daly offered Hearst the ownership of a new property, the Ontario silver mine, in Utah. The syndicate financed the operation, and under Daly's expert guidance the Ontario went on to become one of the great silver mines of the West, making millionaires of all four. For Hearst, Haggin and Tevis it was the beginning of a mining and financial empire. Their next major investment was championed by Hearst. It was the Homestake gold mine in Lead, South Dakota. The syndicate had assumed control of the Homestake properties only two years after the nearby massacre of Custer and his men. South Dakota in those days was no place for the faint-hearted. Hearst himself would not recoil from intimidation and underhand practices if they would serve his ends, and he eventually ended up with control of a broad tranche of leases covering the whole of the massive Homestake lode.[13] The property would soon become the largest gold mine in the United States. It would help Hearst buy the *San Francisco Examiner* and use its influence to win election as one of California's representatives in the US Senate.

Daly, meanwhile, had been active in other fields. Backed by the Walker brothers, another successful mining syndicate, he developed the Alice silver mine in a remote area of Montana, a northern state abutting the Canadian border. Already a thriving gold and silver community, it had been left marooned after the economic panic of 1873, when two railroads that would have connected it to the outside world were abandoned. Ores and metal had to be dragged by wagon across mountainous roads to the nearest rail depots, 150 miles away. The richness of the Alice proved attractive, despite its isolation. As more prospectors came to try their luck, an area to its south began to attract attention. It was hilly district known as Butte (French for small hill). By the time the rail finally arrived, Butte had become a thriving and important silver district. Marcus Daly had secured his own part in the action by purchasing the Anaconda lease and convincing the Hearst syndicate to finance the construction there of a mine and mill. After about a year of prosperous operation, the silver ores began to wane and the Anaconda looked like being one of the syndicate's less successful bets. Instead, however, of simply depleting into barren rock, the news came from the underground workings that a new ore had been found. It was copper sulphide. The great Anaconda copper mountain had been discovered.

To many the West was still a silver- and gold-mining area. The lower-value industrial metals were by-products that could be sold at a profit if supported by a sufficient production of precious metal. The Anaconda copper was poor in silver and possessed only a medium grade of copper. It did not look especially promising. Yet, with the railroad now in place, Daly was convinced of the potential that lay underneath the hill. He needed only to win the support of the syndicate. Here he nearly failed. Tevis and Hearst saw no value in risking a fortune to develop a copper deposit in a remote region far from the main markets. They would need mining on a scale unknown in the West, and their partially smelted copper would have to be railed and shipped across huge distances to be fed into the refineries of Swansea or the East coast. No-one could conceive of a viable refining complex in the West. The partners' scepticism was strengthened when they considered that the copper industry was already dominated by giants such as Calumet & Hecla and Rio Tinto, to say nothing of the Chilean mines that still held the lion's share of world production. It was Haggin who decided the issue. Silently considering the arguments put forward by Daly, he made his pronouncement. The mine would go ahead and he would 'see Daly through on the deal'.[14] The combined determination of Daly and Haggin resulted in a mining and smelting complex that, when finally completed in 1884, was described in glowing terms by the industry press. A fourth force had entered the ranks of world copper suppliers, and it was plain to all that soon Anaconda would be a rival to the Spanish, Michigan and Chilean incumbents. When James Douglas, backed by the Phelps Dodge company, completed the Copper Queen development at Bisbee in Arizona,[15] the question on everyone's lips was 'where would all the copper go?'

THE SECRETAN AFFAIR

The Great Lakes accounted for nearly four-fifths of American copper, but there was unease in Michigan. In a letter to their New York agents, Calumet & Hecla stated bluntly that they should let their buyers know that the Michigan producers intended to retain their present business and 'not to allow themselves to be supplanted by Arizona'. As it turned out, Arizona was to be the least of their troubles. In 1885 the full force of Anaconda swept onto the market. In that year Michigan's share of American copper sales declined by half. The Lakes cartel wavered, barely able to withstand the sledgehammer blows of competition from the giant mines that rose out of the West. But it held. The Boston financiers regrouped in the face of the Western threat. Under the leadership of Calumet & Hecla they forced the price down even further. It reached eleven cents per pound in the spring of 1886,[16] half the price of a few years earlier. It would take more than a price war to beat the likes of George Hearst and James Haggin. Production from Butte continued to climb, and wages were held steady. Yet none of the producers could sustain such ruinous prices for long. The first casualties were the Arizona mines. They shut down in the summer, and towards the end of that year even Anaconda had laid off most of its workforce. Victorious, the Michigan pool now had to face reality. It could maintain the price at this low level and keep the Western concerns closed, but in doing so they themselves would surely also fail. Prices began to climb. Soon Montana was once more producing more copper than Michigan.

The price war had had some beneficial effects. It had expanded demand for the metal. For a time copper had become cheaper than iron. But even the rapidly industrializing urban centres of the United States could not swallow all the output from Michigan, Montana and Arizona. It began to find its way in large quantities onto the export market. The United States became an exporter on a scale to match Chile. Anaconda copper was railed to the West coast, from where it made its way to the H. H. Vivian Works at Swansea or the East coast refineries. Much of it was used in Europe. The Spanish and Chilean producers now saw their markets directly threatened, and the rampant price-cutting for them spelt ruin. Increasingly, the answer was seen to lie in bringing to a halt such destructive competition. Rio Tinto and Calumet & Hecla, already leaders of their own cartels, were no strangers to the idea. Into this receptive environment stepped Monsieur Secretan, with his scheme for a world copper cartel.

Secretan cut a dashing figure in Parisian society, with his expensive suits and flamboyant gestures. He had got his start supplying brass artillery cartridges to the French army during the Franco-Prussian War of 1870–71, and had progressed to the position of director of the Société des Metaux, the leading buyer of metals in France. When he formulated his plan for a copper cartel he was already happily operating a tin cartel. The success of that venture gave him confidence, and now his idea was to form a syndicate that would underwrite the purchase of the entire copper production of the world's large producers. He would corner global copper. The syndicate comprised the Comptoir d'Escompte (one of the largest banks in France), the Paris branch of the Rothschilds and several other French financial institutions.[17] Throughout 1887 agreements were concluded with Anaconda, Calumet & Hecla and the Spanish producers. James Douglas resolutely kept Phelps Dodge aloof from the project. As the scheme lumbered into life, prices jumped sharply upwards. There followed a frenzy of trading in the shares of copper mines, the rush led by Secretan and his friends, as they sought to gain not only from the price hikes but also from the appreciation of the shares.

The weakness of the scheme was that Secretan had undertaken to buy all the output from the big producers. Naturally enough, therefore, throughout 1888 their output increased, and a small mountain of copper began to pile up in the warehouses of the Société. To make matters worse, the high prices were encouraging people to look for alternatives to this newly expensive metal. Some began to search for scrap, and some began to substitute iron or zinc. Others stopped buying altogether. Production continued unabated, and in January 1889 it was estimated that the syndicate held one year's global supply of copper. Rumours were now rife of an imminent collapse. A month later it all came crashing down. Prices fell by half in two weeks. The syndicate panicked and sought to dump the copper at sellout prices. Now the producers stepped in. They threatened massive retaliation if dumping was to take place. The copper stockpile stayed where it was. Its gradual liquidation was administered by the Rothschild bank.

The Secretan scheme turned out to be a windfall for the producers. As Douglas put it, '[our] debt evaporated like dew'. It was the average Frenchman with his savings locked up in a bank who bore the losses. Indeed, only intervention by its government saved France from being dragged into a general economic disaster. As for Secretan himself, he was thrown in prison. One of his colleagues committed suicide. Two other casualties of the period would be those long-time pillars of the industry – the Swansea smelters and the mines of Norte Chico. Initially the Butte and Arizona ores had been roughly smelted on location and then shipped from San Francisco around Cape Horn to the docks of Swansea. The town remained the home of the world's best smeltermen. It also retained its old advantage of cheap coal and cheap labour, both of which were scarce in the isolated mining camps. Smelting and refining, even when the costs of shipping were taken into account, could be more cheaply done in Swansea than in the West. Two technical advances would soon demolish that advantage and bring to an end the 150-year career of the Associated Smelters.

The first was Bessemer smelting. The advantages were obvious of adopting to copper Bessemer's air-blown steelmaking process. It could accomplish in a few hours what it took the Swansea smelters four or five days to achieve. The potential for savings was enormous. For decades all attempts had proved fruitless to adapt the process properly. The trouble was that, unlike iron that reacts violently with oxygen, the speed of reaction for copper was much slower. As the reaction proceeded, therefore, the copper would cool down and cake over the air injectors, eventually bringing the process to a halt. It was no insoluble problem, but progress was agonizingly slow. Perhaps the Swansea smelters discouraged the innovation. Liberating the smelting process from the need for huge amounts of coal would seriously undermine their geographical advantages. The breakthrough was achieved by Pierre Manhes,[18] a metallurgist from the industrial city of Lyon. It was seized on by the Parrot works in Butte. When the mighty Anaconda switched across to the new technique, Bessemer conversion was well on its way to becoming the standard practice throughout the Western copper-fields.

The second innovation was electrolytic refining. The ores of the western United States were not well-suited to the Welsh process. They contained a good portion of silver and gold, metals that the Welsh process was poorly adapted to recover. Small technical improvements had made some headway on the problem, but the real advance occurred when James Elkington patented his technique for the electrolytic refining of copper.[19] The basics of this new process had long been known in concept. It had been the difficulty and cost of generating sufficient electrical power that had stood in the way of commercial application. Elkington established the first industrial electrolytic refining operation at his smelting works at Pembrey in Wales. Then James Douglas followed suit at Phoenixville near Philadelphia. The quality from both was poor, due to the lack of a strong and reliable electric generator, but the recovery of precious metal encouraged perseverance. The challenge was to build electrical generators of sufficient size. It is at this point we should digress a moment to examine the development of electrical power.

The phenomenon of electricity had long been the subject of laboratory experimentation. The Englishmen Humphrey Davy and Michael Faraday had made considerable advances in demonstrating its properties. Faraday, in particular, had advanced the understanding of electrolysis. It took some time for the first practical application of electrical power to appear. It came with the development of the telegraph. In 1843, after years of lobbying from Samuel Morse and others, the US Congress voted funds to build a telegraph line from Baltimore to Washington, DC. By then a British system had been in operation for six years. But it was across the wide open spaces of the United States that the telegraph really showed its value, and soon the continent was criss-crossed by the insulated copper lines. Then a submarine telegraph cable was laid across the Atlantic. The new communications system needed relatively little electrical power. This was fortunate, as most electric generators still could barely rival the power of a steam engine or even a swiftly flowing stream. This began to change when Zenobe Gramme unveiled his new dynamo at the Vienna Electrical Exposition. Then along came Thomas Alva Edison. It would be he who would finally propel the world headlong into the age of electricity. Thoroughness, patience and persistence were his greatest strengths, as well as an indomitable self-confidence that kept him going when others would have long given up. 'Genius is 99 per cent perspiration and 1 per cent inspiration', he told an interviewer later in life. More acidly, the physicist Nikola Tesla remarked that if someone asked Edison to locate a needle in a haystack he would 'with the feverish diligence of the bee, examine straw after straw until he found the object of his search'.[20]

In 1878 Edison, already a rich man, joined the race to develop the electric light. His first burnt cotton filament bulb was promising, and the second glowed for 40 hours. Crowds visited his laboratory in Menlo Park, New York, in order to see these new electric lights illuminating the street. The invention caught on. No less than five companies were preparing to establish commercial electric lighting networks in the city. Most of them were content to keep their networks confined to supplying a single building or factory. Edison decided to construct a much wider network powered by a single central power station. To this end he took advantage of the new large dynamos being manufactured in the workshops of George Westinghouse. Always mindful of the need for publicity, as customers for his network Edison wanted a mixture of residences, factories and banking establishments. The area he chose was bounded by Wall Street and New York's East River and the central generating station was situated on Pearl Street.[21] It is the Pearl Street Power Station that has acquired fame as the birthplace of the modern electrical generation and distribution system. Pearl Street showed that reliable generators of large size had finally arrived.

It was one of those happy conjunctions of technical breakthrough and market demand. Only electrolytically refined copper was of sufficient purity for wiring the new breed of electric motors, and electrolytic refining was the ideal technology for treating the Western copper. Within a year five large refineries had been built in the United States and Great Britain. Four were built where they could take advantage of cheap labour and coal – in Baltimore, Boston, Swansea and New York.[22] Only the owners of the Boston and Montana complex at Butte chose to take advantage of hydro-electric power. They constructed their refinery at Great Falls in the north of Montana. Soon after, Calumet & Hecla would follow suit, constructing the world's largest refinery at Niagara Falls. The scale of these installations was such that electric wiring fell dramatically in cost, thereby expanding demand. By the turn of the century, two-thirds of global copper production was devoted to electric wiring. Thus the year 1891 marks the true takeoff of the electrical industry. It is also as good a year as any to date the death of the Welsh process. It was unable to compete against the much more effective combination of Bessemerization and electrolytic refining.

The year would also serve tolerably well to denote the end of Chile's first copper era. Of all the important copper regions of the world, the mines of northern Chile had been the least prepared for the bruising battles of the era. Its companies had remained small in size and their mode of operation traditional. The difference in scale is illustrated by the fact that Anaconda accounted for as much production as the twenty largest Chilean mines, each separately owned and managed. The richest Chilean ores were now depleted and, while there was plenty of copper in the ground, no individual was capable of raising sufficient capital to invest in the large-scale works needed to make the lower-grade deposits economic. This was not due so much to lack of capital. European and American finance was quite capable of moving into the furthest flung corners of the globe. The real issue has been expressed thus: 'Chile had a mining policy of state revenue, the United States one of economic growth'.[23] The move from Spanish colonial mining laws to the modern pro-capitalist laws had been slow. Mining was still seen as an activity that should be carried out by the gentleman farmer. Thus Chile remained stuck in a centuries-old tradition of small-scale labour-intensive mining. The world was changing. Successful mining and smelting were now being done on a massive scale. Cut adrift in the wild seas of the new global market, the ship that was Chilean copper foundered and eventually sank beneath the waves.

The loss of its major export industry would have prompted many governments into action. Indeed the incumbent government did plan a series of reforms, but it was removed after one of the periodic Latin American political upheavals. The government that replaced it saw little need for action. Why should they, when their coffers were now being filled by a great mineral bonanza in the Atacama desert? In that waterless wasteland, in the border region separating Chile and Peru, endless quantities of nitrates had been discovered. Nitrate was the richest fertilizer then available, as well as being an essential material in the manufacture of explosives. The presence of mineral wealth of this order inflamed what was already a tense territorial dispute.[24] So began the War of the Pacific. It would end in victory for Chile over a disorganized Peru and Bolivia. After the war, each year hundreds of thousands of tonnes of Chilean nitrates were dug up and loaded on ships bound for ports around the globe. They nourished wheat-fields in Australia and beet-fields in France, and, most importantly, the body politic within Chile. The urgency to find a fix for the copper problem was gone. It was left for future generations to solve.

THE TRUST MOVEMENT

The Secretan Affair may be viewed as one particularly violent upheaval in a global revolution. It was a flawed solution to a problem with which industrialists the world over were grappling. That was the problem of economies of scale. The nub of the matter lay in the fact that a machine that could process, say, ten tonnes of ore per hour now cost only a little more to buy than a similar machine that could process only half as much. Whereas before, the technology did not exist to produce the larger machine – motors could not be made powerful enough, the required fittings were too unwieldy, and so on – now the combined advances in technology overcame such problems. Naturally, the tendency was to buy the larger machine. It allowed the cost of production to fall. Or, to put it another way, there were economies in scale. Factories grew in size. The cost of producing each unit fell, and as costs fell prices fell with them. As goods became cheaper, more people could afford them. The era of mass consumption was born. Demand for common items grew at a startling rate. Between the late 1880s and the late 1920s the world's consumption of copper rose eightfold.[25] The period has come to be known as the second industrial revolution.

It was built on the solid foundations of the railway system. It was the railways that underpinned the economies of scale by allowing a steady flow of raw materials to feed the giant factories, and an equally steady flow of cheap grains and livestock to feed the masses who kept the factories turning.[26] They also provided the distribution mechanism for the factory produce as it flowed out into the wider world. More than this, the railways themselves provided the market for a good part of the industrial output. It was not just sentiment that prompted the steel titan, Andrew Carnegie, to name his flagship steel mill the J. Edgar Thomson Works, after the Pennsylvania railroad king. The railroads also required an entirely new form of management. By their nature they demanded a strict adherence to standards and procedures, for without them there would be either frequent collisions or serious inefficiency. The entrepreneurs who built and ran the railroads thus found it necessary to develop a highly-organized form of management to ensure their efficient operation. The more forward-looking factory managers like Carnegie followed suit. The railroad, however, provided only the foundation for the new era. The superstructure was provided by the two new phenomena of chemistry and electricity. From these flowed a torrent of new ideas, processes and inventions. It was a flowering of creativity across the length and breadth of Europe and the Americas that by the eve of the Great War of 1914–18 would result in the industrial world of today being clearly visible. The United States, with its growing population and vast natural resources, seemed often to be the greatest beneficiary of the advances. One example of this can be found in the development of explosives. Having worked with his father and brothers in a St Petersburg armaments factory a young Swede, Alfred Nobel, became obsessed with taming the destructive force of a laboratory substance named nitroglycerine.[27] The final product of his efforts, dynamite (a principal ingredient of which was supplied by the nitrates of the Atacama desert), would provide hard-rock miners of the world with a blasting agent much safer and far more powerful than the time-honoured black powder. It found its largest market in the mines of western America.

Nowhere were advances in chemistry and electricity synthesized more effectively than in the creation of the means for the large-scale manufacture of aluminium. The metal had been known for many years, but the difficulty of freeing it from its tight bonds with oxygen had prevented anyone from obtaining it in any more than small amounts. Its potential was widely suspected. Indeed its strength and lightness had prompted Emperor Napoleon III of France to demand the metal in quantities sufficient to equip his troops. Kaiser Wilhelm II was of a similar view. Unfortunately, the most that could be managed was the fabrication of a lightweight helmet for the Kaiser, a shiny rattle for the Emperor's son and some sets of cutlery to grace the tables of the European elite. The problem was that aluminium stubbornly resisted conventional smelting techniques that are, in essence, the way that most naturally occurring metals are freed from their chemical bonding to oxygen. Metallurgists began to conclude that electrochemical refining was the only answer. Real progress, therefore, awaited the invention of a powerful electrical dynamo, and when Zenobe Gramme's dynamo came to market progress began to accelerate.

A process was developed in Cleveland, Ohio, whereby copper–aluminium alloys could be produced with the help of a strong electric current. Then in 1886 came the decisive breakthrough. In that year the process for the electrolytic production of pure aluminium was patented. The patent was submitted independently by two men in their early twenties: Paul Héroult, a metallurgist working in a laboratory in the south of France, and Charles Hall, a science graduate working in a woodshed in rural Ohio.[28] By 1890 smelters were in operation in Switzerland, Germany and the United States. Cheap electricity was essential to low-cost production, and so the smelters were built to take advantage of hydro-electric sources. The Alps and Niagara Falls became the main production centres for the new metal. Along with the expected military applications and a healthy growth in aluminium kitchen utensils, the lightweight metal proved to be a good conductor of electricity. Soon it was competing with copper to supply the network of electrical transmission lines that was beginning to obscure the urban skies. If the cost of production could be made to fall further, who knew what its limits were? After all, aluminium was the fourth most common element to be found in the earth's crust. A slightly breathless article appeared in an edition of the New York *Sun*: 'Next, the Aluminum Age'.

With developments such as this, the 1890s was a decade of brutal competition. It seemed there was always a surfeit of supply. One cause of this was the number of newcomers wanting their piece of the action. If these entrepreneurs wished to survive, the essential thing was to be cost-competitive. This meant building a large modern plant, running at full capacity and keeping an eagle eye on the costs. Competition consisted in undercutting the other fellow. Was it not Carnegie who said that he would always be the first to cut prices because to be the second meant you had to go even lower? The cost cuts often came from the pockets of the workers, and strikes became a feature of the urban landscape. The most notorious incident of the period occurred when a lock-out in 1892 at Carnegie's Homestead steel works in Pennsylvania led to bloodshed. It began when hired guards from the 'detective' firm of Pinkerton attempted to secure the factory by landing at night from barges floated up the river.[29] Picketing workers spied the flotilla and a gunfight erupted between the men on the banks and the barge-borne thugs. Within an hour two dozen were dead. The mayhem outraged the common folk of Pittsburgh, and for awhile public acrimony was directed at Carnegie and his lieutenants. Public opinion swung his way when an anarchist shot and severely wounded Henry Frick, Carnegie's right-hand man. From his sickbed Frick continued directing his company's response to the strike. It must have been dispiriting for the strikers. Some months later they returned to work, their union destroyed.

Men like Carnegie rode through these troubles and prospered mightily in the market free-for-all. They were quick, bold, aggressive and exceptionally good at finding buyers for their products. For many others, however, business profits could be absent for long periods, even in the good times. It is not surprising that they began to look favourably on the idea of combining with their competitors. The first combinations that were formed broke down, especially where there was no single dominant figure to keep the others in line. Over time the realization grew that a more secure and binding solution was needed for controlling these arrangements. So began the Trust movement. Just as the railroads had led the way in establishing the management model for large enterprises, they also were to lead the way in introducing this movement into American business. In this they had the help of a rising figure in the New York banking world, John Pierpont Morgan. The first industrialist successfully to build a Trust hailed from the new business of petroleum.

Of all the capitalists of America's era of industrialization, perhaps none was better suited than John D. Rockefeller to the task of coldly slaughtering his business opponents. Blessed with an eye for figures that missed nothing – he once took to task a plant manager for a discrepancy of $2 buried deep in his accounts – he drove costs down and tirelessly undercut rivals, buying them out when they were defeated. After fifteen years of tooth-and-claw battling for market share, even he, it seems, had tired of the savage competition in the oil industry. Though he won every battle, the oil-refining business was a many-headed hydra. No sooner would one group be seen off than another would rise to take its place. At last he gathered together a dozen of his largest competitors and convinced them to pool their production facilities under the one banner. No doubt his proposal was delivered in the form of mailed fist in velvet glove. Ownership of the various assets would remain unchanged, but the management of the assets would be assigned, or entrusted, to a single group of like-minded men. There was never any question of which management group that would be. Thus was created the Standard Oil Trust. It was led by John D. Rockefeller, his brother William and eight close associates, mostly employees of Standard Oil. This group would become known, not with much affection, as the Standard Oil gang. Rockefeller's Trust set the mould for the rest of American industry. By 1890 there were nearly 100 of them. They covered the sugar and tobacco industries as well as beef, electrical appliances, nails, whiskey and many others.[30] When at last the Ohio Supreme Court ruled that the Standard Oil Trust was unlawful, Rockefeller merely dissolved it and distributed its assets to companies in different states. In 1899 the monolith was reconstituted in the more friendly legal environs of New Jersey. That year saw the peak of the Trust movement. It was not an edifying spectacle. The movement was as much about market manipulation, price fixing and Wall Street promotions as it was about creating more efficient and stable industrial enterprises. The men who led it would become known as the Robber Barons.

Andrew Carnegie's industrial empire had by this time become the dominant force in American steel. The immigrant Scot's philosophy had always been to drive costs down, win the big contracts by price-cutting, seek out still more orders and push capacity to the limit. When that limit had been reached, he would build another works with the newest technology. He built the United States' first Bessemer steel works and, having been the superintendent of the Pennsylvania Railroad, copied from that company its disciplined management procedures. In addition to the J. Edgar Thomson and Homestead works, Carnegie operated several other steel mills in Pittsburgh, and still more in neighbouring towns. Having secured ownership of the best coking coal in Pennsylvania he took on its owner, Henry Frick, as his commander-in-chief.[31] It was a wise choice. Frick was a tough and energetic manager, one who gave organization and form to what was then a collection of well-run but more or less independent assets. It was he who convinced a reluctant Carnegie to move upstream into control of Minnesota's iron ore. Carnegie had never been one for Trusts, and something akin to panic seized the industry when two of his executives, Charles Schwab and Elbert Gary, began talking about building massive new works that would drive costs down even further.

There were few men in America who could stand up to Carnegie. The threatened steelmakers jumped at the chance, therefore, when one of those few, John Pierpont Morgan, stepped in with a proposal of his own. Morgan had moved from reorganizing the railroads to bringing together industrial companies, all with the aim of reducing the price competition that was raging between them. He had taken Edison Power & Light and its largest competitor and established the General Electric Company. Now he would do the same for steel. He organized the frightened industrialists into the Federal Steel Company and set the stage for a battle royal with Carnegie. In the end it came to nothing. Instead, Carnegie had decided it was time to retire, and in 1901 Morgan announced the formation of the United States Steel Corporation. The new company included

Carnegie Steel, Federal Steel and many other smaller concerns. It was much the biggest Trust of the era, and brought into the one giant conglomerate the activities of iron and coal mining, smelting and the manufacture of steel plate, rails and nails. In all, nearly 800 mines, furnaces and manufactories were brought under the one huge umbrella.[32] Carnegie, now worth about $250 million, withdrew from the field, following John D. Rockefeller into a retirement in which both men would devote their remaining years to philanthropy. One wonders whether they managed to save their souls. For his part, Morgan was far from finished. In 1902 he formed the International Nickel Company, thus securing for US Steel a dominant position in the manufacture of nickel steel plate, a material that was fast becoming a staple of the world's armaments industry.

The story of nickel takes us to Canada, which until the British North America Act of 1867 – the Act of Dominion – had been a collection of British colonies. Though encompassing a vast territory, most of the population was clustered around the northern bank of the St Lawrence seaway and in the land to the north of the American state of Maine. Writers of Canadian history tend to stress that Canadian political unity was a marriage of convenience. Each signatory to the Act of Dominion, they point out, was motivated not so much by the ideals of nationhood but rather by the fact that they had something to gain. While this hardly makes Canada unique among nations formed within the last 200 years, what does make Canada unique is that the inclusion of its western provinces depended on the promise to build a national transport system. Leaders of the populous region of Ontario (which included what is now Quebec) had undertaken to the inhabitants of distant British Columbia that they would send a railroad across thousands of miles of uninhabited prairie for the sole purpose of providing a transport link east for the handful of trappers and fishermen who lived there. To many the construction of the Canadian Pacific Railroad was an act of folly. James Douglas from his vantage point in Arizona described the idea as the inspiration of 'a genius, a knave or a fool'.[33] As it turned out, the railroad was a truly nation-building project, bringing into cultivation the fertile expanses of Alberta, Manitoba and Saskatchewan and creating a highway to the

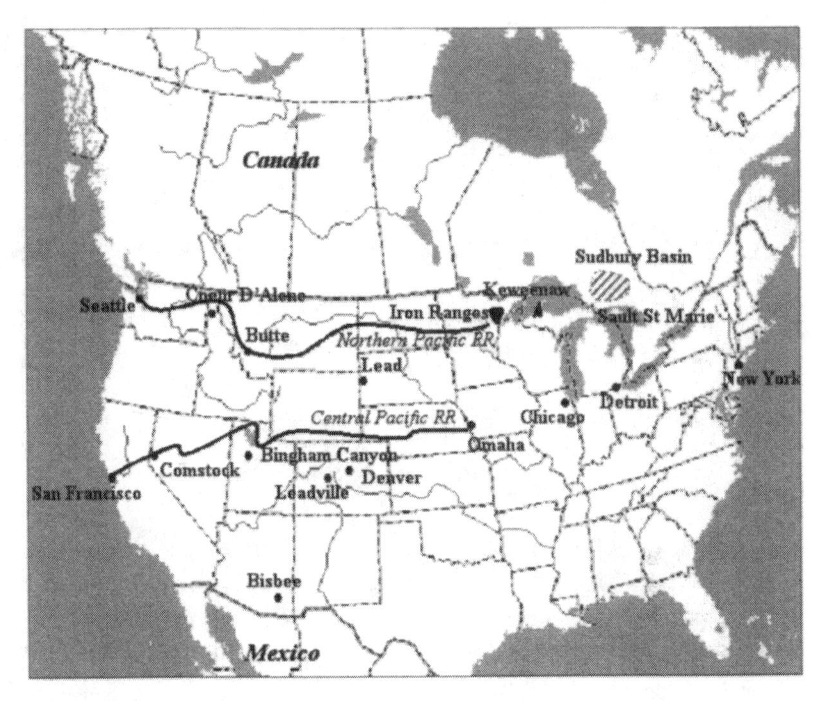

7 North America, 1900

Pacific and trade with Asia. By chance it also led directly to the development of the mineral wealth of the Sudbury Basin.

That there were minerals in this region had been sometimes suspected. For one thing, there was the 'long finger-like Keweenaw Peninsula that pointed out from Michigan like a signpost'.[34] Along the northern edge of Lake Superior, lumps of copper would occasionally be found. For a short while a copper mine had been worked in the area. Little of any lasting importance happened, however, until workers on the Canadian Pacific Railroad exposed a rich ore near the new-built railroad town of Sudbury. It attracted a group of prospectors, who, in the space of twelve months, had located the outcrops of a dozen large copper deposits. Just two of the prospectors, James Frood and F. C. Crean, accounted for five of these finds. Into this bonanza came a chemist working in the iron mines of Minnesota. Backed by a group of Ohio businessmen, Samuel Ritchie purchased as many of the claims as he could get his hands on. He grouped them together into the Canadian Copper Company and began development of the most likely looking prospect, the Copper Cliff. The company and mine property names tell us what sort of deposits Ritchie thought he had secured. We can only imagine his reaction when, having sent a first shipment of Copper Cliff ore to the Orford smelter in New Jersey, the news came back that while copper was certainly present, so too was nickel.

It is doubtful that he was pleased. Nickel had a dubious reputation. The sheltermen of Agricola's time had occasionally encountered it, and had christened it *kupfernickel* (the Devil's copper) because of the trouble they experienced in producing good-quality copper in its presence. In the nineteenth century the metal found some limited uses. One was in the coating of coinage to prevent their tarnishing. Nevertheless, demand remained low because of the expense and difficulty of its production. When large deposits were discovered on the Pacific island of New Caledonia, the Paris branch of the Rothschilds' bank had taken control, and by the time Ritchie created his company, the island was producing virtually all the global output. The Rothschilds used their connections to seek a wider range of applications for the metal and a cheaper production process. As a result of their efforts, a paper on the subject of 'Alloys of Nickel and Steel' was published. The paper spelled out the toughening effect of adding nickel to standard steel.[35] The point was not lost upon the US Secretary of the Navy. For some time naval men had been worrying about the development of armour-penetrating shells. Nickel steel promised to provide an effective defence against these missiles. $1 million of Federal government money, a large sum for the time, was devoted to testing its possibilities. The trials were a success – nickel steel was indeed a formidable defensive material. Once its benefits had been proven, there remained only the question of getting hold of sufficient nickel.

It is here that Ritchie, instead of cursing his luck, began to count his blessings. For the owner of the Orford smelter was none other than Colonel R. M. Thompson, a former naval officer and now a successful speculator and businessman. He was also a friend of the Naval Secretary. Intense work began at

Orford on developing a low cost method of extracting the nickel from the copper.[36] By 1892 they had found it. From this point the alliance of Orford and Canadian Copper began to gain presence in the market, although by this time tensions between Thompson and Ritchie had resulted in the ousting of the latter. Thompson secured the contract to supply nickel to the US Navy, and on the strength of this the mine and smelter works expanded and ventured onto the international scene. Here he encountered the Rothschilds, who were benefiting from the same sort of military supply contracts in Europe. A three-year price war ensued. The battleground was confined to European markets, as high tariff walls prevented the Rothschilds from competing realistically in the United States. Given this advantage and their lower production cost, the struggle ended in victory for the Orford syndicate. When that test of strength was over, the major producers concentrated on reaching a mutually advantageous division of the markets. Another metals cartel was born.

The big test of nickel plate came during the Spanish American War of 1898. The two crucial naval battles of this conflict had each ended in complete victory for the United States. Whilst virtually destroying the Spanish fleet, the Americans lost not a single ship and, indeed, suffered scarcely any casualties. Whilst sceptics may have attributed that success to other factors, the one-sidedness of the battles was seen as proof of the effectiveness of nickel steel. Thus the demand for nickel entered a new phase. It was this that prompted J. P. Morgan, fearful lest the profits from this burgeoning sector went to others, to make his play for control of Canadian nickel. He found in Thompson a man with whom he could do business. Together they formed International Nickel. Thompson was chairman, but US Steel executives dominated the company. Their intentions were clear from the start. The *Canadian Mining Review*, hardly a radical journal, summed it up: 'The International Nickel Company, that was organized recently in New Jersey, is the result of plans to consolidate and control the nickel production of the world'.[37] In other sectors of the metals industry, another man was hatching similar plans.

H. H. ROGERS MAKES HIS MOVE

Henry Huttleston Rogers joined Standard Oil in 1874. He was a coldly intelligent and ruthless operator, a man well-suited to the Rockefeller business style. He rose quickly through the company ranks, gaining wealth and prestige. On Wall Street he was known as the Hell Hound. Living in the shadow of the great John D. Rockefeller could not have been easy for the ambitious Rogers. Casting about for something he could call his own, his baleful gaze fell on the American mining and smelting industries. The Trust movement had not yet penetrated the base metals sector. Here was a goose ready for the plucking. In 1896 Rogers made his move – attacking at once the copper industry and the sprawling silver- and lead-smelting concerns.

It was never Rogers's intention to use Standard Oil money to establish his place in the unfamiliar world of metals. For one thing, John D. Rockefeller frowned on involvement in the unreliable mining industry. Instead, Rogers would use his own large but not inexhaustible funds. To attain his ends, therefore, he needed allies with money and experience. He found them in Alfred Lewisohn, a New York copper merchant, and the Boston capitalist A. S. Bigelow. Lewisohn was a principal in the New York-based United Metals Selling Corporation, and both were major shareholders in the Michigan mines. The three of them, along with William Rockefeller, set about to effect a consolidation of the Butte mining concerns and to forge an alliance between them and the Lakes Pool. Their plan came together with pleasing speed, and in 1899 the last piece of the puzzle, the mighty Anaconda, fell into the syndicate's hands.[38] When Rogers made his bid for Anaconda he found a surprisingly receptive audience. George Hearst was dead (his fortune ended up in the hands of his unreliable son, the future newspaper magnate William Randolph Hearst). Tevis, too, was dead, and both Haggin and Daly had come to see the force of argument behind consolidation in Butte and throughout the copper industry. Haggin sold his stake outright, going on to make yet another fortune in Mexican mines, while Daly threw in his lot with Rogers.

Although they were receptive, this did not mean that the

owners were going to give the property away. In order to afford the purchase price, Rogers and his associates puffed up the prospectus and sold stock in the new Trust at prices well above what was justifiable. The sale of 'watered' stock was an old trick, but this time they went a little far. Their actions led to one of the share market scandals that were typical of the time. The scandal had the usual outcome. That is, no action was taken to bring the perpetrators to book. The Butte concerns were grouped together into the Amalgamated Copper Company, with Daly at the helm. Rogers also formed the Copper Exporters Incorporated that combined the Butte and Lakes groupings in fixing the export price. Together the two companies were known as the Copper Trust.

Not content with tackling the copper interests, Rogers opened up a second front. This time he was aiming at the lead- and silver-smelting interests that stretched from New England, through the Missouri lead-fields and, passing via Colorado, to the far West. Silver was in the doldrums, and many of the beleaguered silver men were only too happy to find the shelter of a Trust. In 1899 Rogers and friends announced the formation of the American Smelting & Refining Company, soon known as Asarco. Others would call it the Smelter Trust. With this success following hard on the heels of the Copper Trust, it seemed to many that Henry H. Rogers held the copper, lead and silver industries of the United States in the palm of his hand. He looked like becoming, after all, the Rockefeller of American metals. Within five years, however, the Hell Hound's dreams were dashed. Two families played their part in his downfall: the Heinzes and the Guggenheims.

The Heinze story is the less uplifting of the two. Stripped to its bare essentials, it is the tale of how a well-connected and talented young lawyer, F. A. 'Fritz' Heinze, managed, with the help of his brothers, to manipulate the Law of the Apex to his own advantage and amass a fortune from the coffers of the Amalgamated Company.[39] It was a battle of wits between one of the mightiest companies in the land and a plucky but amoral challenger – a six-year farce of bought judges, rigged elections, outlandish lawsuits and vituperative duels between captive newspapers. Montana was an isolated and sparsely populated state in which a stage show of this kind could develop into its true magnificence. Heinze's escapades unfolded in the context of a wider political battle. Marcus Daly was now a US Senator as well as Chairman of Amalgamated. His bitter rival for control of Montana was another mining man, William Clark, one of the pioneers of Butte. Vote-buying was raised to an art form in the struggle. Clark, of course, delighted in the antics of Heinze and lent him all the support of his political machine. At its best the spectacle reached truly world-class heights of absurdity, such as the time when Heinze managed to locate a piece of unclaimed land lying between the main claims of the Anaconda lode. The land occupied no more space than would a small room. By claiming that this piece of ground held the apex to the entire Butte deposit, Heinze was granted by a friendly judge an injunction that brought to a standstill all activity in the largest copper concern in the world. The previously ignored field of mine geology was accorded a new respect, as precise definitions of geological structures now became of vital importance.

Like all good shows there was an element of drama. Heinze liked to portray himself as the defender of the downtrodden against the depredations of the mighty Trusts. And in Standard Oil there was enough evidence of depredations to lend him some credibility. To many in Butte he was a crusading hero. But Heinze's show was really all about money. In the end he accepted a $10 million settlement of his spurious claims and left town, only to lose it all in one of the Wall Street crashes. A brain haemorrhage carried him off at the age of 40. The story was a grander-scale version of the goings-on that were commonplace in the Western mining fields. But it had a more serious outcome than most. Quite apart from the money involved, the battle with Heinze so distracted the Rogers team that their larger ambitions were fatally undermined. The Copper Trust collapsed. By then Rogers had also lost control of the Smelter Trust.

THE GUGGENHEIMS

The name of America's greatest mining family is immortalized in the architectural masterpiece that houses one of New York's most famous collections of modern art. The Guggenheim Museum was designed by Frank Lloyd Wright and constructed in 1931. The fortune that built it belonged to Solomon Guggenheim, one of seven sons of Meyer Guggenheim, a Jewish merchant turned smelter owner. The family first came to national prominence when, only a year after Henry Rogers launched Asarco in 1899, they turned the tables on him and grasped control of the Smelter Trust. It was only fifteen years earlier that chance had led Meyer to revive a couple of failing mines far up in the Rocky Mountains, in the bustling town of Leadville, Colorado.[40]

When the elder Guggenheim first arrived in Leadville it was the biggest and busiest mining town in the whole of the American West. Which probably made it the premier mining camp in the world. Colorado had seen its share of excitement in the early years after the California rushes. There had been a disappointing rush to Pikes Peak, and in the years that followed a small but determined industry had grown up in the rough valleys, devoted to smelting the difficult gold ores found thereabouts. The industry struggled to overcome its isolation. When at last there were enough ores being dug to justify the construction of a local smelter, it was Edwin Harrison, a St Louis businessman, who grasped the nettle. His action sparked off a rush to the area around the new Harrison works, and soon it could boast a population large enough to justify a town. And so Leadville was born. Its first mayor was one Horace Tabor, a local merchant and grub-staker who had lived in the Rockies for years.[41]

The practice of grub-staking was common enough in all the mining regions. A local storekeeper would provide some prospectors with food, equipment and whiskey sufficient for a month or so, on the proviso that any discoveries made would be shared according to some agreed formula. Tabor had been grub-staking the itinerant prospectors for nearly as long as he had been resident in the mountains. In 1877 his policy paid off. In November of that year, George Hook and August Rische, armed with Tabor shovels, found what came to be called the Little Pittsburgh mine. It held a fabulous treasure of silver and lead, made Tabor's fortune and caused a stampede into the area.

Other mines opened up. The following year Tabor struck it rich again, this time after having purchased a salted claim, the Chrysolite, from a notorious conman named Chicken Bill. What Chicken Bill did not know was that had he dug just a few feet further he would never have had need of salting claims again. The Chrysolite was nearly as rich as the Little Pittsburgh. Tabor's finds were matched by others, and soon Leadville was booming. It replaced the Comstock as the largest silver producer in America.

Even in a boom town there are many more mines that fail than prosper, and Leadville was no exception. A couple of the town's older shafts, the A.Y. and the Minnie, were struggling. These had been partly financed by Meyer Guggenheim, who, at the time, was a successful Philadelphia lace merchant. When the mines began to fail, their ownership fell to him. He seized the opportunity with both hands. For a time it seemed he would be just another speculator who had poured money down a worthless hole. But when a rich strike was made at last in the A.Y. mine, Meyer Guggenheim was on his way. In the next few years he worked his Rocky Mountain bonanza for all it was worth, all the while bringing his seven sons – Isaac, Daniel, Solomon, Murry, Benjamin, Simon and William – into the business (his two daughters were not considered suitable for participation in the family concern). He made a habit of lecturing them, as legend has it, with homespun wisdom. On the subject of hard work and initiative, for example: 'a roasted pigeon does not fly into one's mouth'. On family unity: 'single sticks are easily broken, but bound together they can resist whatever force is brought to bear.'[42]

The wisdom may have been corny and his dress the traditional garb of an orthodox Jew, but there was nothing parochial or small about Meyer's ambition, or his iron strength of will. Driven on by a burning desire to see his sons succeed, he began to look about him at the wider world. His gaze came to rest on the smelting industry. Here, he concluded, was the real business to be in. Smelters did not bear the risks and disappointments of mining, yet here they were managing to eat away most of his profits! In partnership with a Denver-based smelterman, Meyer Guggenheim formed the Philadelphia Smelting & Refining

Company and built a new silver and lead smelter in the Colorado town of Pueblo. It took only months before he realized that the smelting business was not quite so straightforward as it looked from the outside. After two years the Pueblo smelter was spilling red ink all over the once impeccable ledgers of the Guggenheim concerns. It may have been the end, before it even started, of a great dynasty, had not good luck intervened. The US Congress passed a bill authorizing the government purchase of large quantities of silver, thereby boosting the ailing silver price. The Colorado revival was swift, and the Guggenheims were saved. Meyer did not intend to have to be saved twice.

The progress of the business of Meyer and sons was rapid after that near-death experience. Prompted by a new tariff that prevented the import into the United States of Mexican ores, Meyer determined to move into Mexico to buy some mines and build a few smelters. It was a bold move, but a far less risky one than it seemed to the many Americans whose picture of Mexico was one of bandits, grinding poverty and corruption. A decade before Meyer's move south, an army general named Porfirio Diaz had emerged from Mexico's 50 years of civil war to re-establish order. His 35-year tenure would become known as the *Pax Porfiriani*. Needing to breath some life into Mexico's prostrate economy, Porfirio embarked on a determined effort to attract the foreigner and his life-giving investments. He pushed through the Mine Tax Law that reduced taxes and removed the last vestiges of Spanish colonial intervention in mineral matters. Foreign money came flowing back into the mineral fields to such a degree that when Daniel Guggenheim arrived in Mexico City he found himself in a country in the midst of a mining boom. The practice of the time was for aspiring investors to pay a visit to the President and secure what we might nowadays call 'fast track' approval for acquisitions and developments. Daniel Guggenheim walked from the Presidential offices with concessions for smelters and mine developments. It was the start of a long association with that volatile republic to the south of the Rio Grande. Daniel and his brothers, under the watchful eye of Meyer, built their interests up until they were the owners of not only the Colorado lead smelter but also of two in Mexico, a Mexican copper smelter and a copper refinery in

Perth Amboy, an industrial city just south of New York. They were now a force to be reckoned with. One estimate has them producing two-fifths of the Mexican lead production and one-fifth of its silver.[43]

Naturally, Rogers had invited the Guggenheims into Asarco, but Meyer had stubbornly refused. Without control of the new Trust, the tough, and now aging, patriarch would never allow it. Control, of course, was something the Trust would not even consider delivering. Already the Colorado and Missouri smelters, as well as many in Mexico, were in their hands. Rogers determined to crush these impudent holdouts, just as Standard Oil had always steamrolled its competitors. Astute observers would have had reason to question the confidence of Rogers and his allies. They had paid a very high price to induce the larger independent smelters into their combination, and as a result the Smelter Trust was ill-equipped to sustain a long campaign of attrition against a determined competitor. The price war that now began quickly became too much for the company coffers. With his backers reluctant to contribute any more, Rogers was forced to turn to the Guggenheims. It was a desperate move, and the family showed no mercy. Rogers, Lewisohn and Bigelow resigned from the Asarco board and the remaining directors were given little choice but to welcome the addition of five of the Guggenheim brothers, with Daniel as chairman.[44] The victory left the Guggenheims in control of American lead and silver.

It also left them in control, if the claims of the Western Federation of Miners could be believed, of the government of Colorado. The election of Simon Guggenheim as representative of Colorado in the US Senate was seen merely as confirmation of this. The whole affair elevated the Guggenheims into the first rank of the Robber Barons. For the rest of their careers their every move would be reported in the Press. Few could remain objective. The family would be fawned on by some but vilified by many more. Their next move would be into copper. But before we follow this progression in their fortunes, let us examine a development that was then unfolding and that would end up changing the face of global mining.

By the turn of the century the American West had become a copper country. Far to the south of Butte, the firm of Phelps Dodge had built up a substantial copper presence in Arizona. And in an isolated corner of Utah known as Bingham Canyon, nestled the Boston Consolidated copper mine and smelter. Ben Guggenheim had been offered a stake in the large deposit sitting a little further down the Canyon, but he had declined. Its metal content was absurdly low. The story of Bingham Canyon has many starting points. Let us begin our account with Thomas Greggie, who, in 1887, chanced on some outcropping iron ore as he passed through the remote Mesabi Range on his way to the busy iron mines in Minnesota's north. The find looked promising, and Greggie convinced a few backers to undertake a drilling programme. Only barren holes were sunk, and the drilling company folded amid the knowing nods of the local businessmen. So when the Merritt brothers of Duluth claimed to have found a soft crumbly iron ore in the Mesabi, no one took any notice.

The Merritts, a family of five brothers, were backwoodsmen raised in the rugged forests of Minnesota. Despite some modest success in logging they could claim virtually no experience in business, and none whatsoever in mining. Their find in the Mesabi needed a railroad, and in search of funding Lon Merritt made the journey to Henry Frick's office in Pittsburgh. He was given short shrift. 'Frick did not use me like a gentleman and cut me off short and bulldozed me', Merritt was later to say.[45] At first glance one can understand Frick's reaction, confronted as he was by this unkempt logger and told a tale of vast quantities of pure iron ore just lying on the surface waiting to be scooped up. On reflection, though, Frick's dismissal of Merritt is less understandable. America was still a new land, and this sort of story was being told by unsophisticated prospectors many times every year. Great fortunes were being made by those who were prepared to follow up the leads that chance had dropped in their laps. It was well known that the Great Lakes was an iron region. The legislators of 1854 had done their job well. The steel mills of America had grown up on a diet of Michigan's Marquette

Range shipped through the Sault Sainte Marie. When the Marquette had begun showing signs of depletion, a renewed search had revealed other deposits hidden in the heavily forested ranges on the western edge of Lake Superior – the Menominee, Vermillion and Gogebic ranges. The Mesabi Range lay only a little to the south.

The Merritts would not be ignored for long. Henry Oliver, a once-wealthy speculator now down on his luck, was in need of an opening. Making his way to Duluth, he promised the Merritts financial backing for their mine, wrote them a cheque for $5,000, was accepted as a partner in the venture, and hurried back to Pittsburgh to arrange a loan to cover his cheque. Oliver and the Merritts opened the Mountain Iron Mine in 1892, and others soon followed. It was in the Biwabik mine, which they leased to Peter Kimberley, that first appeared the steam-shovels for which the Range became famous.[46] Steam-shovels were not a new idea. Rail-mounted steam-driven digging equipment was first used by the railway builders, and had been employed here and there in coal mines. Unsuccessful attempts had already been made to use them in the Michigan iron mines. It was, however, in the Mesabi that they first became a standard tool in the miner's armoury.

The soft flat layers of earth lying atop the Mesabi iron ore were ideally suited to their use. A rail spur would be built right up to the rock face, and alongside it was built another. On one sat the shovel, clanking and hissing as it scooped the earth and dropped it into the rail wagons sitting alongside. Along with the shovels came steam-driven drills for drilling holes in the rock. The holes were then packed with dynamite and their charge detonated. Set to work on breaking and digging the soft earth and iron ore and loading it into wagons, the drills and shovels soon proved their worth. With their dozen or so attendants the machines did away with the army of labourers whose task, in a normal open-cut pit, had been to fill barrow loads of ore and dump it over a small escarpment into the rail wagon below. As the miners became more skilled in deploying the new apparatus there developed rows of terraces, on each of which sat a shovel and a waiting wagon train. The ease of digging and rich iron content meant that the Mesabi produced cheaper and more valuable ore than any of its rivals in the Lakes region.

The Mesabi now became the backdrop for one of the more curious corporate battles of the era. The two 'antagonists' were Andrew Carnegie and John D. Rockefeller. Rockefeller was first on the scene. Although wary of the boom and bust nature of mining, he had taken a substantial stake in the Merritt ventures. Barely had he done so when the economic depression of 1893 had wiped out the other shareholders and delivered to him control of the mines at a price even he could not refuse. Other mines had fallen into his lap at the same time, and soon he was a major force on the Range. One might have expected this to have been the cue for Carnegie – whose corporate lifeblood was shipped out of Minnesota's iron-fields – to charge in to defend himself against being held to ransom by the ruthless oilman. Yet the Scottish steelmaker looked on the situation with serenity: 'The Massaba is not the last great deposit which Lake Superior has to reveal'. In part this attitude can be explained by an occasion early in his career, when Carnegie had lost money in an iron mine. He had not forgotten the experience: 'If there is any department of business which offers no inducement, it is ore. It has never been very profitable ...'.[47] So when Carnegie was approached by Henry Oliver with an offer to participate in Oliver's now-substantial Mesabi holdings, the steelmaker was dismissive. Not so his hard-headed right-hand man. Frick struck a deal with Oliver, and in a couple of years Carnegie capital had established Oliver Mining as the second largest enterprise on the field.

Next, in a development that observers interpreted as the opening shot in a war for control, Rockefeller invested in a fleet of Great Lakes ore vessels, calling it the Bessemer Steamship Company. The newspapers began to talk about the possibility of a Rockefeller Trust dominating the Lakes, and the industry was abuzz with excited gossip about the coming battle between two of America's industrial titans. Yet even when rumours were reported of a Rockefeller steel works, Carnegie remained unruffled. Indeed, the Rockefeller move struck him as still more reason to keep out of the Great Lakes. Comparing his rival's developing position as being the equivalent of owning the oil pipelines, he concluded that mine owners would now be in an even poorer position. It seems he judged the situation well, for the eventual outcome was anti-climactic – the two men reached an amicable deal. Under its terms Carnegie would assume the leases of all of the Rockefeller properties and in return would pay the oilman a modest royalty and guarantee to transport the ores via Bessemer Steamship. Carnegie was now much the biggest miner on the Lakes. The experience, along with Frick's counsel, seems to have changed his views on mine ownership. Offering his remaining rivals good terms, one by one he gathered their operations into his embrace until he had become the master of the Mesabi. Why the Rockefeller deal? Carnegie credited himself: 'It does my heart good to have got the best of Reckafellow.'[48] More likely, Rockefeller's distrust of the mining game got the better of him. One wonders whether Carnegie realized the extent of his coup. The Iron Ranges became the most profitable part of his empire. By 1902, nearly half of America's steel was being forged from the Mesabi.[49] When the evaluations were completed on the combined properties of United States Steel, it was concluded that fully one half of the giant company's value lay in the steam-shovel mines of Minnesota.

The efficiency of mechanized mining had impressed a young graduate of the Missouri School of Mines. Daniel Jackling had done the usual rounds. He taught chemistry for a couple of years after graduation, then moved to the Colorado goldfields and then to the Mercury gold mine in Utah. It was there that a local mineowner and retired sea captain, Joseph de Lamar, asked him in 1898 to evaluate a copper deposit in Bingham Canyon. The deposit already had a colourful history. Indeed, it was something of a joke in the area. The reason for this was that its discoverer, Enos Wall, insisted that Bingham Canyon was home to a potentially huge copper mine. Others looked at the mountainside and saw only barren quartz rock flecked with a trace of copper. (The mineralogical term for this kind of deposit is porphyry copper.) Nobody could make money out of so little. But Wall persisted, arguing that if the mining and processing were done on a large-enough scale, there was a fortune to be made. Eventually he interested De Lamar, who in turn was sufficiently convinced to call in Jackling to conduct some metallurgical tests. In the event De Lamar's interest fell through, but the experience had got Jackling thinking. Wall, he knew, had a point. The key to unlocking Bingham Canyon was to understand that the building of the railroad, the mill, the town and the power station and of preparing the mine would cost the same, whether 200 tonnes of copper was processed each day or 2,000. Jackling's calculations showed that by adopting the Mesabi technique of massive open-cut earth moving, and by building a large enough processing plant, the economies of scale so achieved could deliver a fantastic profit.[50]

Not everyone agreed – after all, whoever had heard of a 2,000 tonne-per-day mill? The Guggenheim's chief engineer, John Hays Hammond, and the world's most famous mining man, turned up his nose. It was ridiculed by the most respected mining journal of the day. The project languished, and Jackling went north to work in a gold mine in Washington state. All the while he was refining his plans for Bingham Canyon, until at last he secured sufficient backing to put his vision to the test. While he set about designing the mining and milling operations, he assigned Henry Krumb to undertake a programme of drilling the mountain in order to prove what lay beneath. So thoroughly did Krumb perform his task that his system of grid drilling has remained a standard ever since. Now certain of what lay in store, the building began of the giant mill, and of the railroad that would haul the concentrate from the mill to the Garfield smelter on the shores of the Great Salt Lake. The whole project was incorporated as Utah Copper. The shovels started in at the mountain in June 1906. For those unaccustomed to the open pits of the day, the sight must have been awe-inspiring: two-dozen shovels and trains forming concentric semi-circles on the mountainside. Even as the copper began to flow, capital kept pouring into the project. First it was realized that the mill was too small, and then it became apparent that more mining equipment was needed. In the end it proved to be too much of a strain on the wallets of Enos Wall and the Jackling syndicate. They made it known that they needed a new partner. Now Daniel and his brothers seized their chance. In this way was laid the foundation stone of what would become the Guggenheim copper empire.

Bingham Canyon captivated the mining world. Mechanizing the Mesabi was one thing. It was a rich and easy ore that would have paid off under the old-style hand mining. Utah Copper was quite another thing. It showed that the techniques of mass production could make a fortune out of rock that previously would not have attracted much more than a second glance. Daniel Jackling became a Western hero – much to the chagrin of Enos Wall, who raged in his private newspaper against the world's refusal to accord him credit for the history-making development. While the steam-shovels were what lingered in the imagination, no less impressive was the mill in which the fine-grained copper was sorted from the quartz. First the rock was broken between the jaw-like plates of a mechanical crusher, then it was loaded into a revolving steel cylinder packed with iron balls. Here, in a technique pioneered on the Rand, the quartz was ground down into the consistency of fine sand. These fine particles were then dropped onto vibrating tables, across which a thin broad film of water was flowing. On this grooved surface the combined action of the vibrations and the water had the effect of separating the heavy copper particles from the much lighter quartz sand.

This contraption, known as a Wilfley table, was the delicate heart of the mill. It had been invented in 1895 to treat the heavy lead and silver ores of Denver.[51] The Wilfley table was itself an improvement of an earlier design patented in Germany by the eminent chemist Peter von Rittinger. This table, which was 'vibrated' through the action of a rotating camshaft rhythmically thumping against one side, was in its turn a refinement of an old technique dating back at least as far as Agricola. It is today known as gravity separation, and takes advantage of the fact that particles of rock that contain metallic elements are heavier than normal, barren rock. This property lends itself to helping sort one from the other, particularly when the whole mix is washed in water. The Wilfley tables worked well even on Bingham Canyon ore, so long as it could be ground fine enough to separate the particles of copper from the waste rock. The tables were not without their limitations, however. On the other side of the world, in the town of Broken Hill, deep in the dry interior of southern Australia, metallurgists were discovering just what these were.

The Australian continent, when compared to the North American, had proved a disappointment when it came to minerals. The California Gold Rush had been the springboard to the even greater riches of the Rockies. In Australia, the goldfields of Victoria had been the springboard to nothing much. Pastoralists took the lead in occupying the continent's great flat expanses. Still, the 1890s had seen something of a mineral boom. Over in the south-western desert was the Golden Mile of Kalgoorlie. The old goldfields of Victoria were still active, as were a handful further north. On the island of Tasmania, a wonderfully rich copper pyrite lode was gradually being consolidated into a single concern under the leadership of an American metallurgist by the name of Robert Sticht.[52] Sticht had won his spurs in Butte, an experience that would have taught him a thing or two about consolidating mine properties. He also developed a novel smelting operation that drove down his cost of production and allowed great expansion of the works. For a time it was one of the world's larger producers. The dominant mineral field on the Australian landscape was Broken Hill. This rich lode had been discovered by Charles Rasp, a German chemist turned sheep station labourer. Originally the Hill, so-called because its jagged silhouette gave it the appearance of being broken in several places, was worked for its silver.

Rasp's company, the Broken Hill Proprietary, had staked the best spot and grew prosperous on the rich pickings. Soon its prospects looked sufficiently promising to make it worthwhile employing John Patton, superintendent of the Consolidated Virginia at the Comstock.[53] BHP was only the largest of a dozen or so concerns that lined up along the lode. As mining progressed, the silver-rich rock gave way to a more complex 'sulphide' ore containing a mixture of silver, lead and zinc. So intimately interlaced were these minerals that they could only be separated by grinding them so finely that much of the resulting powder simply washed over the tables and was lost in the waste water. Still, the tables managed to recover enough of the lead and silver to make the effort worthwhile. The 'waste' that remained was rich in zinc. Unfortunately it contained so large a proportion of waste rock that it was unsaleable to the zinc smelters of the day, especially when the contaminating presence of lead was considered. The mine owners were loath to throw it away, so they piled it up next to the mills. It was a nuisance to the residents of the town, who had to put up with the fine dust billowing off the heaps every time a gust of wind blew in off the desert. It was more than a nuisance to the mine owners and labourers. The whole viability of the field was threatened if they could not retrieve the metal trapped within.

Zinc was a new addition, comparatively at least, to the ranks of the commonly available metals. Brass is an alloy of copper and zinc, so clearly the element had been in common use for many centuries. However, prior to the industrial era the zinc used in brassmaking was not the metallic variety but instead was added in the form of calamine, a zinc-bearing mineral. While the brass manufacturers would have been happier using zinc metal – the finished brass quality was easier to control – the problem was that obtaining the pure metal was an unusually difficult task. This is because zinc vaporizes in the smelter instead of melting. Thus, when heated, it would simply disappear out of the furnace. The secret, it turned out, was to heat the calamine in a specially shaped, enclosed clay vessel. The heating continued to cause the metal to vaporize but instead of allowing it to escape to the atmosphere the vessel, known as a retort, channelled it into a separate chamber, where it condensed. Archaeological excavations have shown that a large retorting works was in operation in northern India over what appears to have been many centuries, and it is likely that the technique was taken from there and shipped to Europe sometime around 1740.[54] From that time the zinc-smelting industry of Europe began to grow.

Several works were built in Bristol and Swansea, but over time the centre of the industry gravitated towards those old centres of brass-making, Aachen and Liège and surrounding towns. They took their raw material from the calamine-rich Belgian district of Vielle Montagne. After dominating world zinc production for a century or more, by 1900 this source was near depletion, and the large smelting complexes in the area were looking for fresh feed. So it was that the Australian Metal Company, the latest overseas venture of the fast-growing German metals company, Metalgesellschaft, established itself in Broken Hill. After all, business was good. In addition to the high demand for brass fittings of all kinds, there were also brass shell casings by the thousand needed to supply the armies of Europe, and there was the thriving market for galvanized iron sheets. Faced on the one hand with ruin if they failed and on the other with great rewards if they succeeded, various companies on the Broken Hill field began to experiment. One such was the Sulphide Corporation. Its principals backed the ideas of a British metallurgist named Thomas Ashton. His laboratory work had shown that it was possible to apply an electrolytic process to capture the zinc. A refinery was duly built in the confident expectation that the solution had been found. It lasted less than a year before the company admitted failure and converted the works into a conventional lead smelter. Meanwhile, others were heading in an entirely different direction.

When a finely ground mixture of sulphide minerals, such as those found at Broken Hill, and waste rock is agitated in a vat of water and oil, the sulphides tend to float to the surface along with the oil. This results in a separation through *flotation* of the valuable minerals from the waste. That much had been known since 1860, when William Haynes first demonstrated the effect. Haynes's discovery was just one more particle of knowledge extracted from the laboratory experimentation of the period. As no practical application was known, the idea stayed on the laboratory bench. The Elmore brothers, Francis and Alexander, rediscovered this phenomenon some 40 years later when they were dispatched from their father's gold mine in Wales to investigate a new metallurgical treatment that involved using liberal quantities of oil to help recover valuable ores.[55] Francis Elmore took out a patent on the new technique. Next, Alcide Froment, an Italian metallurgist, threw some sulphuric acid into the oily muck. The bubbles that this generated created a mineral-rich scum that collected on the surface. Some months later a Melbourne brewer named Charles Potter, using a sample of Broken Hill ore, found that simply adding acid without any oil achieved the same thing. Potter acknowledged no debt to

Froment. This of itself need not raise our eyebrows. Simultaneous and independent discovery of technical advances is a frequent occurrence. We have as one example the Hall–Héroult process. We can also point to Alexander Graham Bell. He succeeded in patenting his telephone only a matter of hours before a similar patent was submitted. It did look a little suspicious, however, when the general manager of the Broken Hill Proprietary, Guillaume Delprat, also independently, and at about the same time, made the connection of bubbles on sulphide. Certainly Potter felt cheated. Legal action was inconclusive – Delprat, in fact, escaped having to pay damages only by taking advantage of a loophole in the patent law of the time – and the new metallurgical technique became known as the Potter–Delprat process.

The Potter–Delprat process was effective in floating the zinc and lead particles free from the surrounding waste. The waste dumps were now saleable, and in 1903 two 'flotation' plants (by now the letter '*a*' had been dropped from the spelling) were in operation at Broken Hill. Much the larger of the two was in the BHP works. Meanwhile, faraway in Britain, the Elmore brothers continued to tinker with their process and came up with the idea of subjecting the oily sulphide mix to a vacuum. Then three metallurgists formed a company named Minerals Separation.[56] They bought Froment's patent. They also purchased the rights to the patents of Arthur Cattermole. These specified that if flotation was to proceed to best effect, it was necessary to limit the oil addition and also to stir the whole mixture vigorously. The metallurgists at Minerals Separation combined these notions to create a process that needed little acid and much less oil, and instead relied mainly on mechanical agitation to introduce the air bubbles that created the mineral-rich froth.

Attracted by the heaps of fine sulphide just sitting in the sun, protagonists for all these new processes, and others besides, congregated at Broken Hill. The lode became a flotation laboratory. Into this busy scene came an American engineer named Herbert Hoover. At the age of 31, Hoover was already one of the more respected men operating on the international mining scene. His intelligence, ambition and forceful personality made him a man to be reckoned with. He could make things happen. At once grasping the potential for profits in flotation, he secured an option on several million tonnes of the waste heaps. Gathering backing from a couple of Melbourne businessmen as well as from Francis Govett, the respected London-based chairman of several Kalgoorlie gold companies, he registered the venture as the Zinc Corporation. Then he set to work. First he licensed the Potter–Delprat process. When that failed he turned to the Minerals Separation version. Failing again he scrapped the plant and picked the Elmore vacuum process. The Elmore plant worked, at least well enough to earn a profit. Hoover's success encouraged others, and, on the back of flotation, Broken Hill became one of the world's major sources of zinc.

In actual fact, all the processes worked. The real question was which one, if any, worked well enough to justify the investment in building an industrial scale mill. Flotation was a notoriously temperamental process, and the merit of one technique versus another was often a matter of the precise type of ore fed into the tanks as well as the skill of the individual metallurgist. Each process took time to adapt to the range of ores that may emerge from any particular mine. Much is made in our own era of the impatience of capital markets in their demand for return on investment. The average investor in Hoover's day was no more tolerant. Hoover, a man whose reputation rested on his ability to turn a handsome profit, could not afford the time to tinker. He needed results. As a consequence, he probably rejected the Minerals Separation process a little too early. By 1910, with Herbert's brother Theodore now a driving force in the company, Minerals Separation had mastered the art of bulk flotation. The next year Zinc Corporation scrapped the Elmore process and turned once more to Minerals Separation. It was an outstanding success. With this triumph under their belt, it was time for Minerals Separation to take the new technique further afield. A plant was built in Russia and an office opened in San Francisco.

It is here that controversy enters the story. A dispute arose within the Minerals Separation camp. Theodore Hoover fell out with the directors and left the company, taking with him one James Hyde, who had become his ally in the argument. Hyde some months later installed at a lead-zinc operation in Butte the first flotation plant in the United States. The plant had two consequences. First, Minerals Separation sued for patent infringement, an action that marked the beginning of a legal battle that would rage for fifteen years. T. A. Rickard, a leading mining journalist, captured the scene for the *Engineering and Mining Journal*: 'The litigation over the flotation process is a good Kilkenny fight. Everyone who has got a shillelagh is hitting every head in sight. The Elmore people are suing Minerals Separation and the latter are suing back. In fact, almost everybody is suing almost everybody else ...'.[57] Lesley Bradford, BHP's leading expert, was caught up in the maelstrom, as were several companies that forged ahead on their own. Minerals Separation would eventually prevail after taking the case to the US Supreme Court, and in so doing earned for itself the cordial detestation of many in the mining world.

The second consequence of flotation in Butte was that the nearby copper miners could see for themselves this odd process that was causing so much excitement in distant Broken Hill. They had reasons for taking a good look. While Wilfley tables were successful in recovering the fine flecks of porphyry copper, they were not all that efficient. A good deal of the copper was lost. Even the great Anaconda was experiencing difficulties capturing much more than about four-fifths of the available metal in its ores. Flotation was perfectly suited to recovering the finely ground particles of copper. The innovative engineers building the porphyry mines of Arizona seized on the process. Following suit, after only a few years of determined work, the Anaconda metallurgists could report that 95 per cent of their incoming copper was being captured as product.

Development did not stop there. Whereas once serious investigation had been confined to Broken Hill, now metallurgists and mill operators the world over began conducting experiments with the mineral-laden bubbles. A multitude of discoveries were made. More than once the same insights were gained in several mills quite independently. Workers at Broken Hill discovered that the presence in the flotation tank of small quantities of copper had the effect of greatly enhancing the tendency of zinc to float. A few years later John Myers, a metallurgist at a Tennessee mill of the American Zinc Company, found the same thing when he observed that his laboratory results were much inferior to the results he was obtaining in the full-scale plant. He realized that the difference was due to the fact that he was using brass fittings to stir the flotation tanks in the mill, while in the laboratory the fittings he was using were made of wood. It was a short step from there to confirming copper sulphate as a chemical 'activator' for zinc. When other workers discovered that certain chemicals could be used to 'depress' zinc, the stage was set. Armed with this knowledge the metallurgist could now begin to influence how well the different metals floated relative to each other. So was born selective flotation. Broken Hill led the way in its practical application, helped along by the uniquely Australian chemical, eucalyptus oil, which acted both to activate lead and depress zinc.

It would take years to master the art, but with this last step the value of flotation was confirmed. Now it could be used not only on the mill tailings but also on the mill feed, and by 1920 a row of the frothy tanks was an essential weapon in the mill manager's armoury. Writing in 1932, Pope Yeatman, a veteran of the Rand and the Guggenheim ventures, could say:

> The oil flotation process ... has meant more to the copper industry in so far as increased recoveries and reduced costs are concerned, than any other single factor in the last twenty years. [It] has allowed the treatment at a profit of ores of very much lower grade than could possibly have been accomplished with the recoveries made by old gravity methods of concentration.[58]

Yeatman might well have gone on to say that flotation was the final step in a remarkable transformation of the technologies employed in taking ore to metal. In 1880 men hammered holes into the rock, charged them with black powder and ignited them. Gangs of labourers then shovelled the blasted rock into wagons or rail cars for transport to the mill. Once there, the lumpy rock was broken by steam-powered stamps before being separated into ore and waste – by hand for the larger pieces and, for the finer material, in various devices that took advantage of differences in density. Smelting was carried out using the time-consuming and fuel-hungry methods pioneered in Swansea during the early years of the industrial revolution. This whole series of steps was labour intensive, wasteful of ore and the daily tonnages that could be handled were limited. Forty years later it had all changed. Dynamite had replaced the black powder, and mechanical drills had done away with the back-breaking task of hammering. Great steam-shovels scooped up the rock in open pits. Steam-driven rock crushers had pushed aside the stamp mills. Rotating grinding mills filled with steel balls then pounded the pebbles so produced into fine sand, their ability to handle large tonnages helped by mechanical particle classifiers that diverted the coarser sand back into the grinding mills for another go through. Density separation remained important, but flotation was now the heart of the mill. And in the smelting stage, Bessemer conversion and electrolytic refining now allowed the production of pure metal without the repeated steps of yesteryear. It was as rapid and comprehensive a revolution as any in the history of metallurgy.

THE MODERN MINING CORPORATION

Along with the constructive work in mineral development came the market manipulations. Even respected engineers such as Jackling and John Hays Hammond and his assistants – to say nothing of the Guggenheims and their ilk – indulged in all manner of financial shenanigans. Some surfed the speculative waves to prominence. One such was William Boyce Thompson, whose skills in mine promotion through his company, Newmont, eventually brought him into close friendship with Daniel Guggenheim and secured him a stake in Nevada Consolidated. Thompson had made his first serious fortune during a bout of speculation on the Guggenheim's Nipissing silver prospect. This over-hyped 'bonanza' sparked a briefly hysterical rush to purchase its shares. When the share price fell back to earth, the Guggenheims were forced to refund the stricken investors. Thompson had benefited from access to inside information on the prospect and managed to sell out before the crash. So blatant was his opportunism that the Newmont company biography felt it necessary to state that 'this sort of activity was considered not only proper but admirable'.[59] T. A. Rickard, writing at the time, expressed a different perspective: 'the business of mining assumes an aspect that makes faro look respectable and gives to poker the status of a Sunday school pastime.'

Amid it all, the real work of seeking out valuable mineral deposits continued. The triumph that was Bingham Canyon triggered a wave of similar porphyry developments throughout the American South-west. The Guggenheims established the Nevada Consolidated at Ely, Nevada. It was a runaway success. Suddenly they were not only the world's largest producers of silver and lead, they were also copper barons of the first rank. As the best of the Western prospects were one by one developed, some were emboldened to consider prospects in less familiar lands. Once more, Meyer's boys were at the forefront. When, from high in the Alaskan mountains, reports came of a peak woven with seams of copper of a richness far in excess of Anaconda, they were was ready. It was rough country. The Wrangel Range was far off the beaten track, in a frozen and rugged wilderness that made even the deserts of Arizona look hospitable.[60] A nearby glacier was named after an American military man, James Kennecott. It gave its name to the new lode. It was in the development of the Kennecott lode (the slightly different spelling was the result of a clerical error incurred during the process of company registration) that the true capability of Daniel Guggenheim found expression.

From the coast to this range, through 200 miles of canyons, glaciers and mountainsides, had to be laid a railroad, for there was no other way to bring the ore to the smelter. Into this undertaking Daniel sank a good portion of the family fortune, and still the project demanded more. He persuaded J. P. Morgan to form the Alaska Syndicate with him. They purchased coastal steamship lines to manage the sea transport to the Tacoma smelter in Washington state, tracts of forest to provide the timber for railroad construction, and coalfields to provide fuel for their steam trains. This was in addition to the goldfields they were already working in the Yukon. With such a presence in a sparsely populated state they could hardly avoid controversy. Alaskan politics became polarized between those who considered the 'Googs' a blessing and those who decried their every move. In the eyes of this latter group the family had created in Alaska a virtual fiefdom to add to their ownership of Colorado.

Douglas Braden, an American mining engineer, did with the El Teniente deposit what his former employer had done in the Wrangel Range. That is, he wrestled a rich copper ore body from terrain so formidable that it at first seemed a hopeless task. El Teniente was perched high up on cliffs overlooking a canyon through which raged the waters of Chile's Cachopoal river. Snow blanketed the rocky slopes for a good part of the year. Blizzards and avalanches were common. In this unwelcoming environment Braden would build a mine, and, a mile further down the canyon, a mill.[61] Connecting the two were steel cables that carried buckets of ore from one to the other. It was a technique first mastered in the jagged mountains of Colorado. The mill was equipped with the most up-to-date machinery, and was the site of one of the first successful applications of flotation to copper. From the start, however, the project struggled for funds, and Braden had been compelled to compromise. The mill was built too small and its location was too awkward. More money was needed to do the project justice, and at last he decided to sell. The purchaser would come as no surprise. It was Daniel and his brothers who now added one more giant copper producer to their growing collection. Chuquicamata was next. This time the location was the Atacama desert – one of the driest places on earth. In this forbidding wasteland swept by icy winds from the Andes, the nitrate miners had toiled for years. None had taken much notice of the shallow copper workings that lay not far from the railroad. They were not to know that beneath the humble exterior of the green-streaked hill was the largest quantity of copper ore that has ever been found.

The Guggenheims were now the first port of call for anyone with a property to sell or a need for finance. Indeed, Chuquicamata had first been offered them in 1903, but in the days before Bingham Canyon the prospect had seemed a poor one. Technological progress and the increasing confidence of their engineers changed all that. Securing control of the mountain, they formed the Chile Exploration Company, and soon the vast extent of the ore body was known. To add an extra challenge to those posed by the daunting location, the ore was found to be copper sulphate, a mineral for which no commercial extraction process existed. The metallurgy of the day was equal to the task. It so happens that the action of acid on copper sulphate will dissolve the metal out of its lattice of waste rock. All that then remains is to precipitate it with the help of electric current. Testwork conducted in the civilized environs of New Jersey perfected the process. Steam-shovels fresh from digging the Panama Canal were railed up to the waterless plateau, and rows of tanks built where the leaching would take place. A pipeline carried water from the coast. By 1915 the Chile Copper Company was open for business.

It was becoming a little unwieldy. The Guggenheim copper interests now sprawled most of the length of the Americas. They formed a chain of individual companies, many of them large enough to influence political affairs in their region. In Chile, Guggenheim funds had flowed in such quantities that some said that it had been the cause of the *de facto* currency of the west coast of that continent moving from the British pound sterling to the American dollar. In Alaska, coal mines had been taken into Federal parkland rather than be allowed to fall into the hands of the grasping Googs. With such an array of influential companies, each with their own special needs if their share price was to remain in favour, life was simply becoming too complicated. It seemed as if the sons had forgotten the maxim of their father: 'single sticks are easily broken but bound together they can resist whatever force is brought to bear.' Consolidation was needed. For this reason, in 1917, was created the Kennecott Copper Corporation. It stands as one of the first modern mining multinationals, and was easily the largest of them. It must rank also as one of the larger multinational corporations of the time.

In the early years of the twentieth century, mining and smelting companies were among the giants of industry. Of the 50 largest industrial corporations in the world in 1912, ten were firmly rooted in the mining and smelting industries.[62] The world's largest corporation, almost double the size of the second-placed Standard Oil of New Jersey, was United States

Steel. Morgan's profitable sideshow, International Nickel, came in at number 48. In the top 50 also were Asarco and Utah Copper, both belonging to the Guggenheims. The Copper Barons held three more places in the list – Rio Tinto, Anaconda and Phelps Dodge. The last three entries from the mining world could be found in the newly independent Union of South Africa.

FIVE

The Gold Factories

THE REPUBLICS OF SOUTHERN AFRICA

The year 1869 was a relatively uneventful one in Cape Town, the British trading port located at the southern tip of Africa. There was nothing untoward to influence much the rate at which settlers were arriving from Great Britain. For those who did make the journey, the sight that greeted them as their ship neared port was a town nestled between a spacious harbour and the striking, purple-hued escarpments of Table Mountain. The climate was like nothing that could have been encountered back home. Mild and dry for much of the year, it was also dusty, especially when winds blew in from the east laden with the fine soil of the vast pastoral country, the *veld*, which lay behind the coastal cliffs.

Cape Town had been wrested from its Dutch founders during the Napoleonic War and had been a British possession ever since.[1] The British had imposed themselves on a curious and insular society. The dominant class were the pastoralist descendants of earlier settlers from Holland. The Boers, as they were known, were characterized by a steadfast individualism. Having little use for the norms of urban society, whether of the Dutch or English variety, they had over the years pushed out into distant pastures far from the Cape where they could live their lives according to their own strong beliefs. Unlike the settlers who were conquering the North American frontier, the Boers were not much driven by thoughts of material gain. Their lives were dominated by the desire for land, independence and religious salvation. Most were fundamentalist Calvinists, whose beliefs, due to their long isolation from mainstream European society, had missed the softening and mellowing influences of the centuries that followed the Reformation.

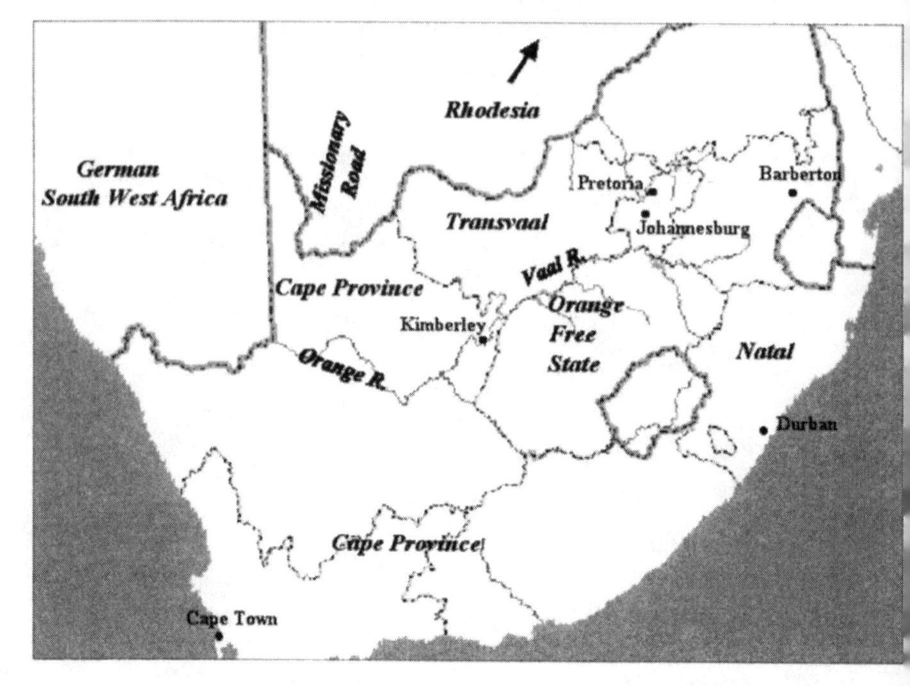

8 South Africa, 1900

It was inevitable that the British and the Boers would clash. The catalyst was the treatment of the natives. The Boers held that the African was inferior to the white man. Enslavement and brutal punishment for minor transgressions were, therefore, permissible. The British, on the contrary, understood that black and white must be equal before the law, and successive British Governors set about enacting reforms to elevate the status of the black inhabitants of the colony. To the Boers this was an intolerable threat to their way of life. Thus they began the Great Trek.[2] It would take these stubborn cattlemen on a long and dangerous journey into rich pastoral areas far to the north-east. There they encountered powerful tribes, whose livelihood also depended on livestock, and a struggle for land began. Boers, British, Zulu and various other local tribes jostled and battled for influence and territory. Alliances shifted with changing circumstances, but by 1869 the political boundaries had crystallized, and southern Africa was divided into four white states.

The largest was the Cape Province, occupying most of the south and west. Squeezed along the north-east coast was Natal, another British colony. Inland lay the two Boer Republics of Transvaal and the Orange Free State. Fractious to the last, the Boers had been unable to confine their differences within the one unified state. Those who held more moderate views and were willing to work with the British tended to stay in the Orange Free State, while the Transvaal had been occupied by the more extreme elements who had continued to drift north. In those days, of course, blacks were denied the franchise, but in this respect they fared little better in the Cape or Natal. The British colonies satisfied both their proclaimed principles and their private bigotry by imposing the high hurdles of literacy and property ownership before a black man was granted the vote.

Economically, the southern Africa of 1869 was heavily reliant on the produce of the land.[3] Wool constituted the chief export. Cotton was cultivated in Natal, and copper ore was exported to Swansea from shallow diggings in the north of the Cape Province. A lone railway stretched from Cape Town to Ladysmith, a distance of a few hundred miles. Adding some spice to the sleepy scene was a new phenomenon. Diamonds had been found on the banks of the Orange River in the ill-defined frontier between the Cape Province and the Orange Free State. Two years later, a bigger find was made in the arid country that lay between the Vaal and the Modder rivers. A new arrival in Cape Town would quickly have heard of this development, because a rush to the diggings had drawn away a good number of the Cape's citizens. Diamonds were the talk of the colony.

THE SCRAMBLE FOR DIAMONDS

The 'rush' is one of the more efficient ways of opening up a new frontier. In the space of a few years a flood of mostly young men will arrive from all corners of the earth. When the rush subsides, the energy and ideas of those who stay are channelled into the less-glamorous work of building a permanent society. It is inevitable that the rush will challenge the existing order of things. Where that order rests on shaky foundations, it will succumb. So it was that in California the Forty-niners swept away the old Mexican society. Where the existing society is more firmly entrenched, as it was in Victoria, the force of the rush will be absorbed, and in the process will create a synthesis of old and new. The experience of the southern African diamond rush, and of the gold rush that followed, illustrates what happens when the existing order rests on foundations of granite. When diamonds were found on the Orange River, the events that ensued had the result of bringing into collision an unstoppable force and an immovable object.

The story of southern Africa's forced march to modernization began on a day in March 1867, when a travelling salesman noticed the children of one his customers playing in the dust with an odd-looking stone.[4] It was a time when throughout the region ignorance of mineral matters was profound. The stone was taken to far-away Grahamstown before it could be confirmed with certainty as a diamond. A handful of prospectors descended on the site, but the excitement soon died away. When more diamonds were found, including a large one now known as the Star of Africa, more people were attracted, and by the end of 1869 a string of tents could be seen stretching for miles along the diamond-bearing banks of the Vaal River. There remained the riddle of the source of the diamonds. The answer came in August of the following year, when one was found on a farm some fifteen miles away. Fortune hunters descended, and soon unearthed other finds. Diamonds were found on the property of the De Beer brothers, and then more were discovered on land close to the De Beer farm, at a place called Blyvooruitzigt.[5] This latter find was called the New Rush. Some years later, Lord Kimberley of the British Foreign Office experienced such difficulty in pronouncing Blyvooruitzigt that the obliging miners renamed it, and eventually the entire region, Kimberley. By now the little cluster of farms was awash with diggers. To picture the space into which they packed themselves, it is necessary to understand something of the physical nature of the diamond deposits. Alluvial deposits of gold and diamonds are ideal for catering to crowds of common men seeking their fortune. The treasure has been deposited by the action of time and water, and is sprawled along river beds and scattered nearby. There is room aplenty. In contrast, the four diamond strikes of the Kimberley occupied each a surface area of no more than a couple of football fields. They were, in fact, the surface of volcanic 'pipes' that pushed far into the earth. The two most easily dug, and the largest, were De Beers and the New Rush. In neither of these did it take long for the diggings to drive deep downwards.

At the beginning, between each claim would be left a thin wall of earth, but as the holes drove deeper, the walls, threatening collapse, disappeared. In their place emerged a landscape of steps. Adjacent claims could differ in level by as much as fifteen yards. It was no place for the unsure of foot. Visitors peering over the side of this huge quarry would shrink back from the sheer drop that greeted them. In later years their view would be obscured by a confusion of wire ropes, each one of which stretched from the top of the pit down into one of the claims far below.[6] Along these wires would be hauled buckets containing ore. The men – mostly black, for the role of the white man lay in the more sensitive business of sorting and selling the diamonds – would climb into and out of the pit on rickety ladders. There was obvious danger and waste in so many independent operations scratching out an existence side by side, so when the restriction on ownership of claims was lifted, a steady process of claim consolidation began. In the years that followed, the Kimberley matured from a mining camp into a well-to-do town, with a stratum of wealthy miners and merchants sitting atop the mass of small-time claim owners, white supervisors and a great crowd of black labourers. Theatres and music halls now acquired a permanent programme, and the accumulation of stone buildings bestowed an air of solidity to what had once been a canvas city. Underlining this was the introduction of steam engines, which began to replace horses for driving the crushing and ore hauling machinery. To many it seemed that the Kimberley had achieved its prosperous destiny.

Such hopes were premature. For some time the diggers had been encountering a tough blue stone mixed in among the soft, crumbly, diamond-bearing gravel. As time went on and the pits sank deeper, it became clear that this blue stone was in fact replacing the gravel.[7] Rumours swept the Kimberley – The blue stone was barren of diamonds! The diamond gravels had given out! Panic gripped the camps. Each day more men sold their holdings for whatever price they could get and departed for the Cape. But not everyone was so ready to admit defeat to this troublesome stone. One of those who determined to stay and fight was the grandson of a rabbi from London's East End. Barnett Isaacs had emigrated to Cape Town following the footsteps of his older brother, Harry. Legend has it that young Barnett had arrived in the Cape with two boxes of cigars as his only source of funds. He wasted no time in joining his brother in Kimberley, where Harry had set himself up as a diamonds dealer. In this vocation the pair prospered mightily, although they would forever be darkly suspected of trafficking in stolen diamonds.

Dealing in diamonds was his day job. By night Barnett Isaacs became Barney Barnato the stage performer or Barney Barnato the prize fighter. His dominant personality, flamboyance and pugilistic prowess soon made him a character to be reckoned with, even in a town as raw as Kimberley. Perhaps influenced by his theatrical bent and his engaging smile, many regarded him as something of a clown. In reality, he was a sharp, shrewd and energetic operator, and not always scrupulous in his business dealings. Barnato had a conviction that the blue stone held diamonds, and plenty of them. He also had the courage of his conviction. Using the money accumulated through diamonds dealing, Harry and Barnato bought some claims in the centre of the Kimberley pipe and started digging into the blue. Soon they could count themselves among the richest men in the Kimberley.

After the blue stone scare optimism returned, but things were different than before. The blue stone might contain plenty of diamonds, but none could deny that it was far more difficult to dig and crush than was the yellow gravel that had lain above it. The realization dawned that successful enterprise now required dynamite for blasting, heavy equipment for hauling the ore out of the ever-deepening pit, and stamp mills for crushing the hard rock. Most realized that it was foolhardy trying to continue to operate alone with a handful of claims. Even the government recognized the inevitable and changed the laws that had formerly restricted the claims that could be owned by any one man or syndicate.[8] A wave of amalgamations and company formation ensued until where there had once been more than 3,000 separate claims being worked, there were now closer to 100 companies. Still, the difficult times did not abate, for there was something far worse than bluestone troubling the Kimberley. Prices remained stubbornly low. Everyone knew that there were too many diamonds flowing onto the market. The solution was no mystery: the number of diamonds being produced must fall. The Brazilian solution – in which the Portuguese had turned the Minas Gerais diamond fields into a heavily policed camp – was obviously not available in a world in which the ideals of individual freedom and free enterprise were held in high regard. These different times demanded a different solution. That solution must be the amalgamation of all claims under a single company.[9] The only question was who and when. In 1885 the railway arrived. With the mood and the stage thus set, a ruthless struggle began that would propel the final consolidation.

THE DIAMOND MAGNATES

The principal antagonists in the battle are well-enough known. Barnato had not rested on his laurels. Through a combination of bluff, charm, sleight of hand, questionable stock-market plays and outright intimidation he manoeuvred himself into control of the largest company on the New Rush pipe, the Kimberley Central. It was a masterful performance, in which Barnato was ably assisted by his three nephews, the Joels, who had made the journey from the streets of London to join their successful uncle. By 1887 the New Rush held only two companies of any size – Kimberley Central, chaired by Barnato, and the Compagnie Française, in which Barnato held many shares but had not secured control. On the De Beers pipe a similar process of consolidation had taken place. Here the driving force was a young man named Cecil John Rhodes. Rhodes had arrived in Natal at the age of seventeen, having been despatched there by his anxious mother in the hope that a stay at his brother's cotton farm would repair his ailing lungs. On his arrival at his brother's farm Cecil discovered that Herbert had decamped to the New Rush, leaving him to mind the farm. This he did, but within a year Herbert had convinced him to make the journey to Kimberley.[10] It is unlikely he needed much convincing. The diamond fields were the only place to be for a young man on the make in southern Africa. Herbert had had the luck to secure a couple of rich claims, and he and Cecil, and soon also their brother Frank, prospered. They prospered further when Cecil arranged a merger of their claims with those of a neighbouring claim holder, Charles Rudd.[11] Leaving Rudd and his brothers to mind the store, Cecil Rhodes set sail for Oxford University. In the years that followed he would make that journey many times.

During his student years, Rhodes, like many of his contemporaries in the diamond fields, came to the realization that consolidation among the claims was the only way to exploit properly the great riches of the Kimberley. Unlike many others, however, Rhodes had the vision to imagine how it could be done and the boldness to pursue his vision. The long voyages between Britain and the Cape were filled with detailed calculations of the costs of achieving, and revenues that might result from, consolidation. In the meantime, Rudd and Rhodes bought up all the claims they could get their hands on. They were not alone in this endeavour, and it was obvious that emerging victorious in the consolidation struggle would be a long haul. While he waited, Rhodes needed more outlets for his energies. After completing his studies, and at the urging of John Merriman, one of the Cape's senior statesmen, he moved into politics, becoming one of the four representatives for Griqualand West in the Cape Parliament. He divided his time between Cape Town and the Kimberley. In the Kimberley he was everywhere. A brilliant and tireless negotiator – 'every man has his price' – he wore the others down with charm, persistence and skill. By 1887 the De Beers pipe was in his hands.

It would be Barnato versus Rhodes. Where Barnato was bold, opportunistic and domineering, Rhodes was persuasive, manipulative and calculating. Where Barnato was backed up by friends similar in outlook to himself, Rhodes had more carefully marshalled his allies. He had the allegiance of the best financial brain in the colony, a diamond buyer by the name of Alfred Beit, a principal in the firm of Wernher, Beit & Company. It was easy for a newcomer to underestimate 'little Alfred'. A diminutive and rather meek appearance masked a skill approaching brilliance at dealmaking and an encyclopaedic knowledge of the diamond fields. Rhodes was always a shrewd judge of others, and the alliance struck between him and Beit was an unusually felicitous partnership. Rhodes supplied the claims and shareholdings, while Beit, with his European banking connections, supplied the funds.[12] And funds were sorely needed as a bidding war began for control of the shares not already in the hands of the two antagonists. The entry into the fray on the side of Rhodes of the Rothschild banking empire was a severe blow to Barnato's hopes of victory.

If any family deserves the title of the modern Fugger it is the Rothschild family. From their origins in the Frankfurt ghetto they had risen to prominence by acting as moneylenders to several German princes. The Revolutionary and Napoleonic Wars that together lasted from the early 1790s to Waterloo in 1815 provided their great opportunity. In their early stages Nathan Rothschild set up office in London and pushed his family concern into becoming the channel through which loans and payments were smuggled through Napoleon's blockade to assist the British war effort. British victory assured the future of the family, and over the next few decades they became financial advisors to heads of state across Europe. Their loans played a significant part in bringing into being Europe's rail system. Along with this came the power to influence the course of events. The Paris branch was able to advance loans to the French government after the disaster of the Franco-Prussian War of 1870–71, thereby possibly averting revolution, and certainly saving the government. When the British Prime Minister, Benjamin Disraeli, finally got his opportunity, in 1875, to buy a stake in the Suez Canal, it was to the Rothschilds he turned for the huge loan he needed.[13] There was even some Fugger-style involvement in minerals. As security against a loan to the Spanish government, the family had assumed control of the Almaden mines. By 1870 they were the richest family in the world. Paris, London and Vienna were their principal offices, with Paris over time becoming much the largest. But their influence was on the wane. The Rothschilds had made their money mostly by brokering the sale of government bonds. As the financial markets became more sophisticated and the middle classes began to deposit their surplus cash in large deposit banks, governments began to bypass the private banks.

This change coincided with the move of the Rothschilds into the financing of mining concerns. Large-scale colonial mine developments were (to an extent) the railways of their day, demanding plenty of money and being intimately connected with the political aspirations of European governments. The Rothschild family concerns acquired large stakes in a Venezuelan gold mine, a Mexican copper mine and a clutch of Mexican lead mines. They controlled the nickel production from New Caledonia. In the wake of the Secretan affair the family not only managed the liquidation of the stockpile, it also gained control of the Rio Tinto mines.[14] Later they would become principal financiers to Anaconda and the Amalgamated Copper group. They had even held an interest in a loss-making diamond company in the Kimberley, and through this experience had concluded the necessity of amalgamation on the field. To assist in understanding all of this unfamiliar terrain they established a consortium with a group of like-minded London financiers. It was called the Exploration Company. Under the leadership of Hamilton Smith, an English mining engineer, the Company's brief was to seek out likely mineral prospects throughout the world and promote them on the London Stock Exchange, always offering first refusal to the consortium.

Given the position of the Rothschilds in the mining world, we are justified in suspecting that when Cecil Rhodes recruited Gardner Williams as the manager of the De Beers mines, he considered more than just Williams's management skill. For Williams was the South African representative of the Exploration Company. With Beit providing contacts with the Continental banks, including the Paris branch of Rothschilds, and now Williams supplying the connection to the London branch, Rhodes had positioned himself for gaining access to virtually unlimited funds. He made full use of the opportunity, and eventually overwhelmed Barnato through the sheer size of his bank account. The battle ended at last in the early hours of 28 February 1888, with an agreement signed by Cecil Rhodes, Barney Barnato, Alfred Beit and Solomon Joel. The De Beers Consolidated Mines Limited, ever after known simply as De Beers, had assumed control over most of the diamond wealth of the Kimberley. As its Chairman, Cecil Rhodes stood at the helm of the global diamond industry. At the age of 35, he was young for such a lofty position. As for the Kimberley itself, with the signing of the cheque it passed rapidly into conservative middle age. The romantic field of diamonds soon became an efficient diamond factory.

THE SCENE SHIFTS TO THE RAND

The amalgamations on the Kimberley fields had pushed out many of the small claimholders. Disinclined to follow a life of tending cattle or working for a wage in one of the larger diamond companies, many of those dispossessed struck out to seek their fortune in fields unknown. Gold was uppermost in their mind, and the Transvaal was rumoured to be a gold country. In 1883, in the rough mountainous region along the border of the Transvaal and Swaziland, a small strike was made. A richer one occurred the following year. A town named Barberton grew up as the result of the hundreds of diggers who came to the area. Barberton was no place for the lone digger. The gold was held in tough quartz reefs, and heavy stamps were needed to liberate the metal from its bonds of rock. In the event, the capital needed to fund this equipment was surprisingly easy to come by. For in the eyes of many investors, the Transvaal was already favourably associated with the riches of the Kimberley. It had one even greater advantage. The men of the Cape stock exchange had all, as schoolboys, heard of the Queen of Sheba and her fabulous gold mines. Legend, and the vague tales brought back by explorers, hinted that the Queen's great mines had existed somewhere to the north of the Transvaal. No prospectus promoter worth his salt could fumble an opportunity like that. Companies were formed bearing names such as Kimberley Sheba and Kimberley Imperial.[15] Every opening was taken to impress on the investor Barberton's association with the legendary golden treasures of old, and the vast diamond treasures lying a mere stone's throw away.

Funds came pouring in. Dozens of companies were floated on no more than the whiff of gold. Much of the money came directly from the London Stock Exchange, and such was their popularity in London that a special name was accorded the shares of the Barberton companies. They became known as Kaffirs. On the strength of the Kaffir boom, Barberton enjoyed three years in the limelight as the Hope of the Transvaal. The crash came when the Exploration Company did the sensible thing and despatched the Californian gold miner, Gardner Williams, to assess the field properly. Williams's report was damning. The reef, he said, was patchy and did not continue at depth. There was little local water power, and certainly no coal, so that what ore there was had to be hauled long distances before it could be fed into the water-powered stamp mills that had been so laboriously and expensively carted from the Cape. The response was panic. Investors dropped Barberton shares as if they were red-hot coals. In the space of a few months the dusty bustling town became a dusty abandoned town. In the process, South African gold acquired a reputation from which it would not quickly recover.

At the height of the Barberton frenzy, gold was found at a lonely place called the Witwatersrand, a stretch of hilly country lying some 35 miles south of the Transvaal capital of Pretoria. The promising strike was made by the brothers Struben, one of several groups of prospectors who were trying their luck in the area. The gold was contained in a curious large-grained conglomerate that protruded from the farmland lying at the foot of the Witwatersrand range.[16] The early arrivals gave this unusual stone the name of *banket*. With all the excitement at Kimberley and Barberton, the wider world paid little notice, and the Strubens, with some few others, were left alone to make the best of their discoveries. This small band persisted, and in February 1886 the first 'Rand' gold – produce of the Strubens' stamp mill – was sent to Cape Town. By then events were gathering pace. On the Langlaagte farm, less than two miles from where the Strubens worked their stamps, another find was made of the outcropping banket. It was richer than anything that went before it, and the excited prospector dubbed it the Main Reef. It was the golden heart of the Rand. When news of the Langlaagte strike got out it touched off a rush to the area. Many who had missed out on their share of the Kimberley spoils made a dash for it. J. B. Robinson, a once prosperous but now bankrupt diamond miner, led the charge. Speaking fluent Afrikaans, he engaged in a whirlwind of negotiations until he managed to buy, using money loaned him by Alfred Beit, the rights to most of the Langlaagte farm and to other adjacent ones.[17] Scurrilous rumour had it that he had taken advantage in more ways than one of the Langlaagte's owner, the widow Oosthuizen.

As was the case in Barberton, it was clear from the start that the Rand would yield its riches only to those with capital to spend. But from where would the capital come? When Barberton crashed, promoters offering Kaffir investments – Rand, Barberton or otherwise – were *persona non grata* on the Cape Town stock exchange. They had no chance in London. Had not Gardner Williams said of the Rand: 'If I rode over these reefs in America I would not get off my horse to look at them.'[18] He was not alone in taking a dim view of this odd-looking rock with its fine-grained gold. Yet gold there was aplenty. And in neighbouring Kimberley there lay a deep pool of mining experience, financial strength and business acumen. As time wore on, the two drew closer together, and, as they did so, the gold began to flow. In the process the Rand was transformed from farmland to a landscape disfigured by piles of waste rock that occasionally rose high enough to obscure the view of an unbroken line of steam-spitting, smoke-belching mills stretching from west to east. An impressive town, Johannesburg, had grown up in the midst of it all.

The Kimberley magnates were all present and accounted for.[19] There were Cecil Rhodes and Charles Rudd presiding over the grandly titled Gold Fields of South Africa. Alfred Beit and his partner, Julius Wernher, had established their representatives in a fine stone building located on one of Johannesburg's busiest street corners. The Cornerhouse, as it was called, had become the hub of a rapidly expanding network of companies and holdings that spanned the Rand. Even Barney Barnato, by now the richest man in southern Africa, had finally deigned to interest himself in the goldfields he had once dismissed as 'Robinson's cabbage patch'. Arriving in Johannesburg to the welcome of a brass band (hired by his agent), he threw himself into the business of gold. Johannesburg Consolidated Investments, his flagship company, soon took its place among the most important concerns in the region. Barnato, Rhodes and their peers did not have the Rand to themselves. Thanks to the rich surface outcroppings, the fruits of the banket were easy to pick. A band of black labourers, a row of stamp mills and a liberal helping of mercury to capture the fine gold – that was all anyone needed to claim a piece of the action in those years. There was plenty of room for the small-time capitalist.

This happy state of affairs persisted for a good three years. Yet for the astute observer there were warning signs of trouble ahead. With increasing regularity, the ore being hauled up from the bottom of some of the deeper shafts showed a bluish tinge, and the way it glinted in the sunlight revealed that it was pyritic ore. Pyrite ore is stubbornly resistant to yielding its gold when submitted to mercury amalgamation. No matter how much mercury is poured onto it, the gold remains locked in its matrix of rock. Pyrites spelt ruin for the Rand. The full weight of the disaster struck in 1890, when the shafts reached a depth at which the banket turned almost entirely to pyrites. Throughout that year scores of companies collapsed. Johannesburg saw much of its population decamp in despair for the Kimberley. Percy Fitzpatrick, one of Beit's men at the Cornerhouse, summed up the mood: 'Grass will grow in the streets of Johannesburg within a year.'[20] Well might he say that. He was a veteran of Barberton. Yet there were some on the Rand who had a different experience of disasters. The Kimberley had been through its own blue stone crisis. For those who had lived through it, the words of the irrepressible Barney Barnato rang most true: 'The gold is here in the earth beyond a doubt. Money and patience will overcome all difficulties.'[21] But not even Barnato could have guessed how soon those difficulties would be overcome. For a technical solution to the problem had already been found.

John Stewart MacArthur was born in Glasgow. On leaving school at the age of fourteen he secured an apprenticeship in the Glasgow laboratories of the Tharsis Sulphur & Copper Company. It must have seemed a plum job. Tharsis was then at the height of its prosperity and influence as one of the central concerns in the fast-growing British chemical industry. MacArthur's job required him to evaluate the mineral content of the ores that were shipped in from around the world. Mostly his work involved the copper pyrites from the Tharsis mine in southern Spain. No doubt, also, he was asked to evaluate the ores that came from the newly opened gold district of Mysore. Famous from antiquity for its diamonds, when gold had been found in Mysore, Charles Tennant, the longstanding chairman of Tharsis, had invested a substantial sum. The mines had proved much more difficult to work than anyone anticipated, and Tennant more than once found himself in the galling and unaccustomed position of being on the receiving end of harsh words from shareholders. Amongst the most pressing problems associated with Mysore gold was the difficulty of winning it from the quartz rock in which it was found. To the rescue had come Henry Cassell, and his patented chlorination process for dissolving gold out of rock. He certainly said the right things: 'We are going to do for the gold mining industry what Bessemer did for iron and steel.'[22] It must have been one of the few times in his life that Tennant, the worldly-wise Scot, was taken for a ride. The Cassell process was a failure, and the man himself absconded, leaving Tennant with no gold process and a minor financial scandal on his hands. It was then that he turned to the obscure chemist who had laboured so long in his employ.

So it was that John MacArthur took the reins of the Cassell Gold Extraction Company. Things were looking up for him. Longing for a challenge, and no doubt bored with his laboratory routine, he had already teamed up with a pair of Glasgow medical doctors, the Forrests, and the trio had set to work looking to make their fortune in the rapidly expanding world of chemical products. With MacArthur's ascent to the helm of Cassell, the three focused their attention on the problem of

Mysore gold. The subsequent development of the cyanide process has parallels with that of mercury amalgamation. As was the case with mercury on silver, it had long been known that cyanide could dissolve metallic gold. This had been demonstrated by the Swedish chemist Carl Wilhelm Scheele as far back as 1783.[23] In the years that followed that discovery, the phenomenon was studied by a succession of eminent chemists, yet no serious attempts seem to have been made at applying the technique to the recovery of gold from ore. That is, until the failing gold mines of Mysore provided a compelling reason.

MacArthur and the Forrests drew up a list of the possible chemical solvents of gold, and soon found that cyanide was the most effective on ores. That was the easy part. More difficult to solve, and just as important, was the question of how the gold could be retrieved back from the cyanide solution. They discovered that small shavings of metallic zinc could accomplish this task, and in 1886 was born the MacArthur–Forrest process. Essentially it worked by using cyanide to dissolve the gold out of its matrix of rock and then using shavings of zinc to precipitate the dissolved gold out of the cyanide bath. In the world of chemical processing it was simplicity itself. It was also a godsend to the worldwide gold industry. The Cassell Company began to send metallurgists to Australia, New Zealand, India, California and anywhere else where mine owners frustrated by the disappearance of their gold in the mill tailings could be found.[24] When pyrites struck the Rand mills, the MacArthur–Forrest process was primed for the rescue. By the end of that unhappy year, the cyanide mill on the Robinson mine was showing the men of the Rand what could be done. Cassell had been right about one thing – his company had done for gold what Bessemer did for steel. As quickly as it had fled, confidence returned to the Reef.

Up to now all that was known of the banket was that it sloped southwards at a steep angle into the earth. How far down it went before petering out was, in the days before cyanide, a matter of engaging, but essentially idle, speculation. Now that the cyanide process had unlocked the gold that lay at the bottom of the shafts, it inevitably prompted the question: How deep did the banket really go? Only twenty years before, the answer to that question would have been merely guesswork. No one could see through rock. The Comstock miners had prospered or bankrupted depending on whether they were lucky enough to stake a piece of ground under which lay one of the silver-rich raisins. The typical mine manager must have felt each day he was entering a casino. One month the ore was rich, the next it was poor; indeed, it might even have disappeared altogether. Once lost, the gold-bearing seam might be relocated through geological sleuthing or it might be lost forever. Now a new technology was available to change all that – the diamond drill. With it the human eye could pierce deep into the earth.

Deep-level drilling was still a novel technique when, in October 1892, a diamond-tipped drill began to bore into the earth somewhat to the south of the main reef outcrops.[25] It was searching for banket. Hopes began to fade when the drill passed the depth of 500 yards without any show of the gold-bearing ore. When it reached 600 yards, it was beyond the depth at which its performance could be guaranteed by the manufacturers. Still the drill crew continued to drive downwards. Seven hundred yards came and went. Their persistence was rewarded when at nearly 800 yards below the surface, the diamond teeth of the drill cut through the Main Reef. The fragments of banket that rose to the surface were wonderfully rich in gold! Further drilling showed that the reef descended into the earth in an entirely predictable fashion. When its consistency was coupled with the efficiency of the cyanide process, it was if the Rand was home not so much to gold mines but to gold factories.

By now the prospects on the Rand were exciting plenty of interest. Hamilton Smith himself visited to assess the opportunities at first hand. On his return he published an article in *The Times* of London in which he spoke glowingly of the future for the Rand.[26] In the next eighteen months his Exploration Company floated several deep level companies, all of which were backed by Beit and the Rothschild Bank. Following Smith's advice they began buying up the leases to the south of the line of the Main Reef, betting that deep underneath lay a fortune in gold. Beit did not neglect his friend and ally Cecil Rhodes. In this way Beit and Rhodes became the magnates most associated with the early ventures into deep mining on the

Rand. When Rhodes employed the services of the world's foremost mining engineer, John Hays Hammond, it was a clear sign that the pace of development was about to quicken.

The arrival of Hammond can be seen as confirmation of the value and importance of the Transvaal and its banket. He was already a figure of almost legendary achievement. Hammond was born in San Francisco into a prominent military family.[27] His father had played a major role in the Civil War, and throughout his early career Hammond had no difficulty in gaining access to the highest levels of American society including, shortly after his marriage, being the guest of honour at a White House banquet. Choosing mining as his career, he travelled to Germany to spend three years studying at the Freiberg Mining Academy, because in the 1870s there were, so his father said, no American mining schools worthy of the name. As a young graduate he impressed sufficiently the hard-bitten George Hearst to be offered the important position of mill superintendent at the Homestake in South Dakota. He declined the offer, preferring to accompany Clarence King on his pathbreaking geological survey from the Sierra Nevada to the Rockies. The next ten years saw Hammond shot at and ambushed on many more than one occasion as he ventured into some of the wildest corners of Mexico and the American West. He tells the tale of being caught up in a general gunfight in a saloon in the notoriously violent gold town of Bodie, Arizona. Through it all he showed himself a courageous leader, with an uncommon skill for evaluating and managing mining prospects. He became a wealthy man. When Barney Barnato invited him to manage his gold interests, Hammond at first declined. He makes plain in his autobiography – which, it must be said, is remarkable for its bombastic tone – that he never had much faith in Barnato, being especially wary of the latter's reputation as a share manipulator. He accepted a year later, when the Presidential election brought to power the Democrat Grover Cleveland. This, according to Hammond's own account, convinced him that tempestuous economic times lay ahead for the United States.

His relationship with Barnato was a dissatisfying one. And when, six months after arriving in South Africa, he was approached by Cecil Rhodes, he needed little convincing to switch camps. For in Rhodes Hammond had found a man who had a thrilling vision for the future. Unlike Barnato, Rhodes was also willing to let Hammond have his head, and so it was that the American engineer put all his energy and skill into the task of sinking the first deep level shafts. He doubled, then tripled, the rate at which the shafts were being sunk. The Rand looked on in anticipation. With perhaps 100,000 people now living in Johannesburg, and with some of the most eminent mining men making it their home, the Rand could now truly claim its place among the world's premier mineral fields. With prosperity, however, had not come contentment. The miners had created an island of raw and raucous capitalism within a rather backward-looking society of cattle farmers, whose chief recreation was not to swill beer in a public house but to read the Bible aloud to their family. It is little wonder that the Boers looked with suspicion and hostility on these uncouth intruders.

THE AMBITIONS OF CECIL JOHN RHODES

The act that was unfolding on the Rand must be seen within the developments on the wider canvas of the African continent. For centuries the Europeans and Arabs had regarded the continent principally as a source of slaves. Portuguese, French, Dutch and British had all participated in this cruel commerce. The slaves were marched from the mysterious interior to the European coastal settlements, and from there transported in death ships to the Americas. Typically the men were prisoners of war, and the women and children taken from captured villages. Despite the vigour of the trade, the Europeans made no effort to establish colonies on the continent, as they had done in the Americas, the Pacific and in southern Asia. Indeed, until 1875 the Cape Colony was the only European possession of any size on the entire continent. Diseases such as malaria and yellow fever prevented the Europeans from advancing very far inland (unlike their Arab contemporaries, who ranged across the savannah and through the jungles in pursuit of their victims). In any case, few governments saw the point in attempting to subdue hostile tribes when their military and economic gain would be the possession of a few sleepy mosquito-infested outposts doing desultory trade in items of little value. Early in the nineteenth century the European slave trade was abolished. Roughly coinciding with this came the first of a steadily increasing number of European adventurers who began to follow and map the great river systems of the Nile, the Congo, the Zambezi and the Niger.

This new information attracted great interest in the capitals of Europe. It was a time when the idea of empire was gaining currency in the foreign policies of many European governments. Everywhere the thought became more attractive of possessing a colony or two in Africa. The first move came from a most unexpected quarter. In 1883 King Leopold II of Belgium revealed that he had signed treaties with various tribes, and in so doing had secured political hegemony over the vast and fertile Congo basin. The coup stunned the world and led to the Berlin Conference, in which fourteen European states came together to establish ground rules for how the continent of Africa might be exploited.[28] The conference was superficially a success, but none of those present seemed to have treated very seriously its outcomes. Instead, there began what history has come to know as the Scramble for Africa. In a mere fifteen years the entire continent, with the single exception of Ethiopia, lay in the hands of one or other of the European powers.

Cecil Rhodes was one of those Europeans captivated by the imperialist ideal. By the time he had completed his Oxford studies he had developed a naïve, yet sweeping, belief in the superiority of the British people. From where this belief came is a matter for speculation. One influence may have been the lectures of the Oxford historian, William Ruskin. Others maintain that the intellectual depth of Rhodes's convictions owed more to the kind of tub-thumping editorials that could be found in colonial newspapers. The document in which he summed up his beliefs was a will in which, at the advanced age of 24, he bequeathed his entire estate to the British Colonial Secretary. The estate would be used for 'the establishment, promotion and development of a Secret Society, the one aim and object whereof shall be the extension of British rule throughout the world'.[29] This Secret Society was to perfect a system of British colonization that would encompass the entire continent of

Africa, the whole of South America, the Holy Land, the Malay Archipelago, Japan, the seaboard of China … . The list went on. In short, Rhodes viewed the greater part of the globe as populated by 'the most despicable specimens of human beings', who clearly needed the enlightened guidance of the 'finest race in the world'. Assuredly the will was a strange document, yet the sentiments expressed therein would motivate him throughout his life. And Rhodes was never a theorist or a mere dreamer. By the time of his death, at 48, he had shaken southern Africa to its very roots and changed it forever.

His formal involvement in matters political began with his election to the Cape Legislative Assembly as the representative of the Boer-dominated seat of Griqualand West. Two goals defined his political career. The first was to achieve a union of the Boer and British states in southern Africa. The second was to extend British colonial dominion deep into the heart of Africa. His first move was, in 1889, to found the British South Africa Company, the charter of which was to engage in activities to develop the lands to the north of the Transvaal. No northern boundary was set to limit the range of the Company's activities. The significance of the vagueness of the northern boundary did not escape Belgium's King Leopold. By this time Leopold had set up what was in effect a private fiefdom covering the vast area of the Congo basin. It had been christened the Congo Free State, and to it had recently been added a highlands area to the south named Katanga.

Leopold knew that Rhodes was going to make a grab for Katanga, and he could guess why. Katangan mineral wealth seemed to be a prize worth fighting for. After all, a Scottish missionary had not long before published a book on his central African travels in which he described the digging and smelting of the copper ores of the region. Rumour had it that gold was also to be found. While Leopold could be reasonably sure that Rhodes had little real support from the British government, there was no doubt that the jingoistic London press was urging him on. A map appeared in *The Times* that showed Rhodes's British South Africa Company in control of Katanga.[30] Leopold was alarmed. The rules of the Scramble for Africa held that effective control of any area must be backed up by a written agreement with the local tribal chiefs. This meant that a deal had to be struck with a certain Chief M'siri, who wielded a despotic rule over the Katangan plateau. Several Leopold- and Rhodes-funded parties were despatched to M'siri's capital, Bunkeya, to compete for his favour. In the end, Leopold – if anything, a more determined empire-builder even than Rhodes – prevailed. In due course he would establish a company, Union Minière du Haut Katanga (United Minerals of Upper Katanga), to which he would assign the task of developing the mineral wealth of the region. The first Katangan copper was railed to the coast in 1911.[31]

For Rhodes, to be thwarted in the Congo was no doubt galling, but there was plenty of other work to do. He had only a tenuous hold of the fertile grassland immediately to the north of the Transvaal. There lived the powerful Matabele on lands as yet unclaimed by any other European power. To correct this state of affairs he sent Charles Rudd, his faithful partner of old, to the kraal of the Matabele chief, Lobengula. Rudd returned from his trip, which included all the usual hair-raising adventures that accompanied such missions, with a document bearing the seal of the Great Elephant. He had won the mineral rights. The Rudd Concession had in fact been obtained through the kind of shameful trickery of which the settlers of the American West would have been proud. Shameful or not, it was enough for Rhodes. He next assembled and despatched a group of men he named the Pioneer Corps. Their task was to subdue this dangerous land.

They did not fare well. The Matabele had not forgotten Rudd's duplicity, and clashes of increasing ferocity occurred. Rhodes decided to up the ante. Just as when he needed financial advice and help he turned to Alfred Beit, when he needed brave and bold leadership in the face of physical danger he turned to Dr Leander Starr Jameson. It was to Jameson that he now turned. He made the right choice. Jameson launched a series of savage attacks on the Matabele, slaughtering warriors, massacring women and children and reducing their villages to ashes. Soon Matabeleland lay at the mercy of the doctor and his band. The campaign had two results. First, it hastened the formal recognition of Matabeleland and Mashonaland as the British Territory of Rhodesia – a recognition that probably marked the high-point of Rhodes's life. Second, it gave both Jameson and Rhodes great confidence in the value of bold action against the odds. This confidence would lead them both to disaster.

THE JAMESON RAID

The scope of Rhodes's activities in 1895 would have taxed the resources of three or four very capable men. He was Chairman of De Beers, his gold interests were pouring money into deep levels on the Rand, he was Prime Minister of the Cape Colony, prime mover in the creation of white settlement in Rhodesia, and, despite the Katangan setback, was pressing ahead with plans for the construction of a railway from the Cape to Victoria Falls. He was even seriously considering the launch, with Hammond's help, of a new global copper cartel along the lines of the Secretan syndicate. He had London on his side, and most of southern Africa eating out of his hand. It must have been vexing, therefore, to encounter one man whom he simply could not win over. That man was Paul Kruger, the President of the Transvaal.

Kruger has, over the years, been much maligned. His enemies have portrayed him as a pig-headed, corrupt, despotic ignoramus with filthy personal habits. He has even been vilified for being ugly! The truth is somewhat different. The Rand gold industry had placed Kruger in a difficult position. His Treasury badly needed funds to pay for the many tasks involved in creating a viable nation. Thus, reliant as he was on Rand taxation, he was not against the success of the mines. On the other hand, he and his fellow Boers had every reason to distrust the avaricious mob who had pushed their way into their midst. To some of the more fundamentalist of the Boers, it must have seemed that a new Gomorrah had arisen less than a day's ride from their capital. Most of the others simply distrusted the British. They had good reason. In 1878 the British government, seeking to force the pace of South African federation, had annexed the Transvaal. It had taken a military defeat of the British troops – at

Majuba Hill – and the liberal policy of Gladstone to restore their independence. Now they were back, in a different guise to be sure, but their aims were still the same.

Some of the younger and more progressive Boers saw in the gold mines and miners a great opportunity to modernize their country, and to these Kruger would listen from time to time. To the majority of Boers, however, the Rand miners were *uitlanders* (literally 'out-landers'), aliens who could only disrupt their orderly and God-fearing republic. Kruger, as a typical Boer and also as a democratically elected President, adopted a policy in tune with the wishes of his people. That is, he aimed to contain the intruders and limit their scope for causing damage. To this end he refused them the vote and denied them state subsidies for the education of their children. Kruger had a second problem in that he lacked experienced administrators who could collect the taxes and operate public utilities. To solve this problem he adopted the expedient of setting up state monopolies through which much of his taxation was collected. The production and import of dynamite, for example, was a state monopoly with prices to match.

To all of this, many of the mine owners complained loud and long that it was all terribly unfair. In reality, Kruger's taxes in no way matched the abuses commonly wrought in the Robber Baron era in the United States. Nevertheless, among some sections of the miners the idea of rebellion began to take root. It grew throughout 1895, the rebellious passions feeding on themselves and flourishing in the insular Johannesburg community. It must be doubted how serious these notions were. But running in parallel with these popular sentiments, a more serious rebellion was being planned. The leaders were Cecil Rhodes and his ever-faithful ally, Alfred Beit. The plot brewed throughout the second half of the year. A local Johannesburg group was established that would co-ordinate the rebellion. John Hays Hammond and Frank Rhodes were involved, as was George Farrar, a leading mine owner, and the Chairman of the Rand Chamber of Mines, Lionel Phillips.[32] Guns were smuggled into Johannesburg in oil drums. Most ominous of all, in November of that year Dr Jameson set up camp just outside the border. With him was a large body of his Pioneer Corps, fresh from their depredations in Rhodesia. By now the whole Transvaal knew something was up. Jameson, sensing that delay would be fatal, wanted to go in.

Throughout December, crudely encoded telegrams went back and forth. The Johannesburg conspirators urged delay, Jameson urged action. Rhodes and Beit attempted to judge the right moment. Were they encouraged by Joseph Chamberlain, the British Colonial Secretary? No one has ever been able to say for sure. Then came the point when Jameson could stand it no more. On 29 December 1895 he rode across the Transvaal border at the head of 500 mounted troops. His plan was to ride into Johannesburg – or would it be Pretoria? (he hadn't decided) – and in so doing precipitate an armed rising among the miners. His bold action would surely trigger a barrage, before which Kruger's government would crumble and Jameson would win the day. After all, it had worked before. Yet, in the event, the Raid was a total failure. The raiders got a few hundred miles into the Transvaal before being captured after a brief battle. They had ridden into a trap. In Johannesburg the shock was palpable. The conspirators froze up, and days later they were arrested without a shot being fired. Rhodes was forced to resign the Prime Ministership. Chamberlain, far away in London, disavowed the whole affair.

The Jameson Raid has ever after been a subject for controversy. Who was behind it? More importantly, why did they do it? One long-held theory continues to contend that the Raid was an internal revolt of the *uitlanders* against the so-called despotism of the Kruger government. Geoffrey Blainey has advanced the view that the costs and risks inherent in the sinking of the deep levels drove Rhodes and Beit to the conclusion that deep mining could never be economic under the taxation policies of the Kruger regime.[33] Less widely spoken of has been the *hubris* of Rhodes. His long list of accomplishments may well have left him with the feeling that he was invincible. When Kruger refused to join the federated states of South Africa, Rhodes's reaction would have been to swat him aside like a pesky fly. The Raid, haphazardly planned and wretchedly executed, demonstrated Rhodes's contempt for his opposition.

The post-Raid policy of Kruger displayed no such *hubris*. Due legal process had pronounced the death penalty for Hammond, Frank Rhodes, Lionel Phillips and several others. Instead of marching them to the gallows, Kruger commuted the sentences and some months later the conspirators were in Cape Town, having been, in effect, ransomed by Rhodes and Beit. Kruger also made a genuine effort to reform his government's policies toward the gold industry. Some of the monopolies were abolished. The industry itself was in rude health, its output having tripled between 1894 and 1898.[34] In that year it produced one quarter of the world's gold. The owners of the main mining houses were now so rich that they were known in the Press as Randlords. But Kruger's efforts were deemed inadequate. The British High Commissioner to South Africa, Alfred Milner, pressed hard on him to grant the franchise to all white men living in the Transvaal. Kruger knew that to do so was to agree to the end of his republic. It was an impasse that could have been resolved in many ways. But Chamberlain and Milner wanted war. Faced with intolerable provocation, Kruger delivered an ultimatum to the British.[35]

Chamberlain needed no second invitation. Led by their Generals, Botha and Smuts, the Boers put up an unexpectedly vigorous fight, but there was no serious uncertainty about the outcome. The Anglo-Boer war was over by 1902, and the South African republics were united under the banner of the British Empire. Many saw the war as the worst example of imperialist aggression. It inspired the economist and journalist J. A. Hobson to write *Imperialism*, a study in which he characterized the conflict as an example of how capitalists, having glutted their home markets, must resort to building colonial empires to provide an outlet for their surplus goods. This colonial expansion was actively supported by the great banking houses: 'Does anyone seriously suppose that a great European war could be undertaken by any state ... if the House of Rothschild and their connexions set their face against it?'[36] The idea would later be seized on by Lenin.

After the war, the industry moved into a new phase. For one thing there were new leaders in the major mining groups. Rhodes was dead. The master imperialist had made a good show of bouncing back from the setback of the Jameson Raid.

Yet the strain of it would have been great. Never in good health, the hot southern summer of 1902 finished him. He died of a heart attack in January of that year. Barney Barnato, too, was dead. His last years had been plagued by worry, and increasingly by mental illness. The murder by a blackmailer of one of his nephews had added to his paranoia. On a voyage to London he leapt out of his deck-chair and threw himself overboard, drowning before help could reach him. Hammond had departed the scene to join the Guggenheims as their chief consulting engineer. His salary was said to be colossal, although one doubts that it would have been sufficient to match his ego. Robinson was now spending most of his time in London, and Beit was more and more in the background. The men who replaced them were not pathbreakers like their predecessors. Their skills were more suited to the challenge of managing the gold factories. It was a good match. With the war over, the chief concerns of the new men would be the costs of production. It was a task that called for the careful work of professional engineers.

THE PROFESSIONALS

The Rand was home to thousands of Americans, British, French and Germans as well as, of course, scores of thousands of native Africans. This rainbow of nationalities was nothing new in a goldfield. Many were fortune hunters of the kind who had made the gold camps their home, from Nevada City to Bendigo. There was a large contingent of Australians. So many, in fact, that when General Botha's surrender heralded the end of the Boer War, there was celebration from Australia's Broken Hill to the Tasmanian copper town of Lyell.[37] The expatriates on the Rand included a large group of mining and metallurgical engineers. They were part of what was by now a corps of professional mining men spread across the globe. Most were American and British, and the great majority held formal qualifications in mining or metallurgy. They had gained these qualifications at schools that for the most part had been in existence for less than twenty years.

The notion of a formal education in the mineral professions was not new. The first mining school of which there is any surviving record was founded in the old German silver town of Freiberg. It dates from 1702.[38] Schools were later opened in Mexico, Peru, Bohemia, France, Russia and elsewhere. Despite this, for the first two-thirds of the nineteenth century, the idea of formal technical training continued to be regarded by the majority, especially in the English-speaking world, as a conceit. The mine captains of Cornwall had received no such training, and saw no need for it. And if the world's leading hard-rock mining region could not claim even a single school devoted to the field, what possible use could they be? A similar attitude existed across the Atlantic. Cornish captains, German smelter men and their home-grown American counterparts had capably managed the first waves of underground mining and smelting developments in the American West. Only rarely were these men blessed with any formal training.

The maturing of the industrial era brought with it a different attitude to education. The mid-century establishment in London of the Royal School of Mines – which, in its first years, held the title of Government School of Mines and of Sciences applied to the Arts – created the first mining school in the British Isles. Even in tradition-bound Cornwall attitudes were changing. A public meeting was held in the village of Camborne, a location that represented a compromise between the rival towns of Redruth and Truro.[39] There it was decided to found the Miners' Association of Cornwall and Devon. It marked the beginnings of organized education in the region. Fifty years later, in the twilight of Cornish mining, the old jealousies would at last be put aside and the classes held in scattered centres brought together into the Camborne School of Mines. There was a blossoming of American mining schools. Columbia University in New York opened its School in 1864,[40] and by the turn of the century the United States was home to seventeen schools, its closest rivals being Germany and Britain, both of which could claim five. As their graduates prospered, the mining schools received private funding. The busts of Julius Wernher and Alfred Beit flank the entrance to the Royal School of Mines, testament to their help in establishing the imposing building that replaced the former ramshackle arrangement. Herbert Hoover became a noted benefactor of Columbia University, his *alma mata*.

It is tempting to attribute the rise of mining education to the increasing technical demands of the industry. This argument, however, holds little water. The technical demands of the Cornish copper mines at the time of Watt's introduction of his steam engine were surely as great for the miners of the day as were the challenges of Bingham Canyon and the Rand deep levels for the mining men of their day. Yet, in that era, the Cornish turned their back on the European-style schools, preferring to stick with their captains. The rise of mining education was, in fact, due in large part to pressure from outside the industry. Let us examine this a little further.

Since the dawn of the industrial age there had been a steady growth in what we may call professional engineering institutions. Railway engineers had their own association, as did those who designed and managed the construction of ocean-going vessels. Membership of a professional body bestowed a prestige and trustworthiness on the individual. The main conditions of membership were an unblemished character and a demonstrated knowledge of technical matters pertinent to the field. Among this professional engineering family, the miners and metallurgists, leaving aside those involved directly in iron- and steel-making, were the poor cousins from the wrong side of the railway tracks. Many considered the industry to be one of dubious repute, where technical issues were at best of secondary importance.[41] They would have agreed with Mark Twain when he described a gold mine as a hole in the ground 'with a liar standing next to it'. A self-respecting engineer may well offer his services to a mining concern, but few would permit themselves to become identified with the industry.

It was in the last third of the nineteenth century that all this changed. It was driven mostly by the fact that mineral developments had become ravenous consumers of capital. The Spanish copper investments of Tharsis and Rio Tinto and the great copper and silver projects in the American West made large demands of the capital markets in London, Paris and the moneyed cities along the Eastern seaboard of the United States.

Bankers would not long tolerate pouring such sums into an industry that was so poorly managed as to be unable to muster its own professional association. The bankers needed men they could trust – and whose language they spoke – to advise them on mineral prospects in far-off places. It is no accident that mining school graduates in those years frequently worked as roving consultants for city-based investment firms. The task of actually developing and operating the mine was still left to the practical man.

This need of the financiers for trustworthy and technically educated mining men drove the industry towards forging its own professional identity. In 1871 the American Institute of Mining Engineers was launched. Two decades later, similar institutes had been brought into existence in Australia and Great Britain. Through their influence, respect for formal education grew, and by the end of the century the positions of mine and smelter superintendent – once filled only by 'practical men' – were very often the preserve of graduates of the mining schools. Indeed, the situation had moved to such an extent that in 1896 the President of the Institute of Mining and Metallurgy felt able to say: 'it is a notable fact that the most important and responsible positions in mining are going more and more to the scientifically trained man, while the purely practical man ... is gradually assuming a secondary role'.[42]

It was a truly global profession. The continent-hopping career of John Hays Hammond differed from many others only in the magnitude of its success. Where once the German master miner could be found in the leading positions in every mining field in Europe, now the new cadre could be found in every mineral region, from the *velds* of southern Africa to the mosquito-infested forests of equatorial South America. Despite the fact that most had command of several languages, it was an English-speaking industry. A survey of any of the world's main mining centres would have revealed that the majority of the leading technical men – the mining engineers, the geologists and metallurgists – were largely American and British, with a good helping of Australians. The German presence was also strong, especially in metallurgy, a fact that reflected the pioneering efforts of German industry in harnessing the potential of industrial chemistry.

To this new generation fell the task of guiding the industry through the transition from managing mines as if they were casinos to managing mines as if they were factories. Yet, despite the increase in the ranks of graduate engineers, there were still relatively few men who could effectively develop and instil a factory style approach to mining operations. John Hays Hammond was one. When it became necessary to speed up the sinking of the deep level shafts, he set one of his young engineers the task. When the engineer protested that he had no experience in such matters, Hammond replied that that was why he had chosen him for the job.[43] His principal charge was to break up the job into its component parts and examine each part to understand how it could be done more quickly. It was a reply worthy of Henry Ford. Men of Hammond's calibre were not easily found, and on the Rand was pioneered a management technique that aimed at spreading this scarce resource more widely. Here, the large mining houses – now down to about half a dozen substantial companies – established a system by which a core team of senior technical experts would consult each of the operations within the group. It was known as the 'consulting engineer' system. Where mine ownership was not so concentrated, or where the field was smaller, other management systems had perforce to evolve. It is in the goldfields of Kalgoorlie that we find the model of the 'managing consultant'.

THE LAST GREAT RUSH

Of all metals it was gold that most underpinned the international nature of the mining industry. For it remained the frontier metal. It was the dream of every prospector who pushed into the African river systems, the frozen wastes of Siberia and far North America, the South American jungles and the parched expanse of the Australian outback. The dreams were fuelled by a steady stream of successes. The great alluvial finds of the gold-rush era had been replaced by a host of smaller fields in places as far apart as West Africa, central India, the Southwest of the United States and eastern Australia. Then, in the single decade of the 1890s, prospectors stumbled across three major finds – in Australia, the Yukon and at Cripple Creek in Colorado – which, when added to the increase from the Rand, caused global production to jump nearly threefold.

The most spectacular was found deep in the waterless expanse of Western Australia. The timing of the discovery was fortunate. The London capital market was in the grip of gold fever. British bankers, having recovered from the blue stone scare, were pouring large sums into the Rand and had become accustomed to the idea of investing in fine-grained gold deposits in waterless and isolated corners of the Empire. So when promising quartz outcrops were found at a place called Kalgoorlie, interest was immediately sparked, and it was not long before the 'Westralian' stocks were being fought over by British investors eager for a piece of the bonanza. In the intoxicating atmosphere the objective consideration of mineral prospects gave way to wildly optimistic appraisals, and eventually to outright fraudulence. While such dishonesty had its practitioners in Johannesburg and everywhere else, Kalgoorlie was in a league of its own. Arriving in 1897, the young American mining engineer Herbert Hoover was taken aback by the 'rank swindling and charlatan engineering' that pervaded the field.[44] When they woke up to the goings-on, the London-based financiers realized that they needed someone they could trust to bring this unruly yet lucrative field into line. They turned to the impeccably credentialled consulting firm of Bewick, Moreing & Company.

Algernon Moreing and Thomas Bewick were consulting engineers who had built their reputation on appraising mine prospects across the globe. As the years had passed they had begun accepting contracts to develop and operate the prospects on which they had advised. In this they were not alone. The by now venerable firm of John Taylor & Sons ran a clutch of Spanish lead mines as well as virtually controlling the Indian goldfield of Mysore. When the Bewick, Moreing engineers began to arrive in Kalgoorlie, it was only a few years before the firm was running some of the smaller or struggling prospects, most notably the Sons of Gwalia mine that had the 22-year-old Hoover at the helm. He set the style for the future, cleaning out the corruption and putting the management on a firmer footing. The focus switched to lowering costs and increasing the productive life of the mines.[45] The style had a hard edge, and any union agitation was swiftly dealt with. So successful did this approach become that, by 1905, Bewick, Moreing were managing half of the field's production. The influence of the firm, along with engineers and managers imported from Colorado, the Victorian goldfields and the Transvaal, was one reason why Kalgoorlie became for a time the world leader in the processing of fine-grained gold ores. A visitor from the Rand could declare that the Kalgoorlie field was ahead of his home town in the skill of its managers and its willingness to adopt and improve on new techniques. Such could be the influence of good technical training on even the most wild and woolly of distant fields.

As the engineers moved in, there was less room for the independent prospector. In Kalgoorlie these men eventually found themselves pushed by the companies into the waterless scrub, where there was little gold to be found. They could have been forgiven for wishing their campsites had been located in a cool, well-forested land, where water was plentiful and the summers mild. For those sufficiently tempted by such a thought, in 1898 their chance arose. Setting out from Kalgoorlie, they could travel by bullock dray to the port town of Fremantle, and there board a ship bound for the ramshackle Alaskan port of Skagway. Once in Skagway they could join the hordes that had come together to descend upon the Yukon, an isolated river in Canada's frozen North.

The Yukon River drains the melted snows from the western slopes of the mountains of north-west Canada. It continues flowing westwards across the border separating Alaska from Canada. Once on the American side, it picks up several large tributaries before emptying itself into the Bering Sea at a point somewhat to the south of what is now the township of Nome. It was in the Yukon basin that prospecting parties found gold at a place they named Forty-Mile Creek. A town, Dawson, was founded and it grew to be inhabited by a couple of hundred hardy adventurers. The task of gold-digging in this new field was no sinecure. Winters were dangerously cold, and the summers plagued from beginning to end by swarms of mosquitoes and biting flies. For most of the year the gold-bearing gravels

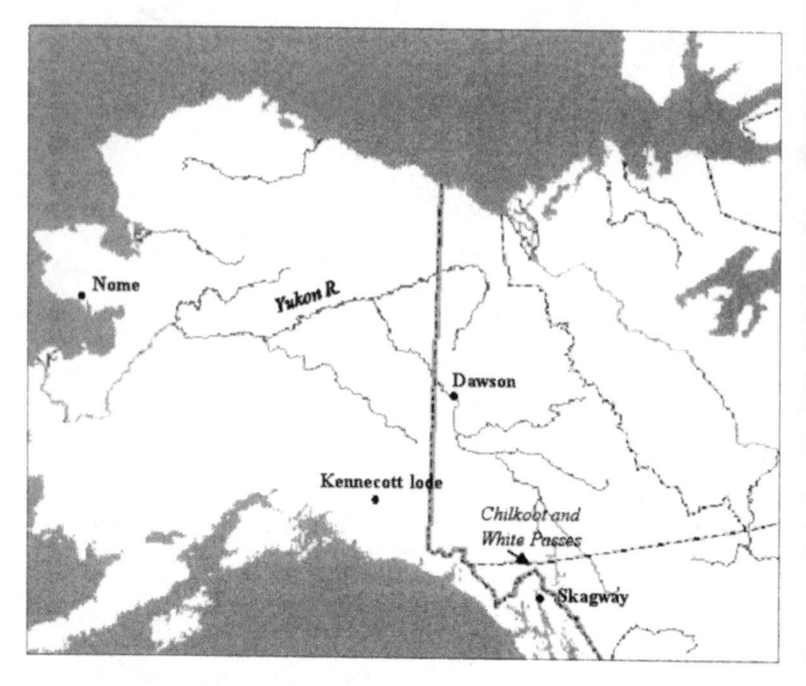

9 The Klondike, 1900

were frozen solid, and digging was possible only with the aid of ground-melting fire. Occasionally a minor find would be made somewhere else along the Yukon, but each was only enough to keep the small local crowd of prospectors interested. Few newcomers were attracted into the region.

The wheels were set in motion to change this sleepy state of affairs when, in June 1896, an Indian named Skookum Jim struck gold on the Thron-diuck River, not far upstream from Dawson. Such was Skookum Jim's haul that the site became known as Bonanza Creek. A rush followed that virtually closed Dawson. In the excitement, somehow the Thron-diuck became the Klondike.[46] Many of those several thousand who rushed upriver to the Klondike in the summer of 1897 would have made their fortunes. Word, of course, got out, and when steamboats loaded with gold began arriving at San Francisco, the rumours were confirmed. It must have seemed like California all over again. A great mass of humanity – estimated by some to be as many as 100,000 – set out for the Klondike to seek wealth and adventure.

Perhaps 35,000 actually made it.[47] Some opted for the steamboat voyage up the Yukon, while others made the circuitous journey along Canada's river systems. The most dramatic route was to disembark in the notorious town of Skagway and cross to the upper reaches of the Yukon via the Chilkoot and White Passes. A striking photo exists of an unbroken line of men and horses battling their way up the steep and icy slopes of the Chilkoot. Once at the top, no man could proceed until the Royal Canadian Mounted Police was able to certify that he had brought with him the requisite 1,000 pounds of provisions. For most this required a three-month period of shuttling back and forth, as not even the strongest could carry such a load up either of the slippery passes in one go. Many turned back when faced with the enormity of the challenge, and not a few of these would have found temporary solace in the brothels and gambling houses of Skagway. Enough of the starters completed the journey to the town of Dawson to ensure that plenty of gold was soon pouring out of the Klondike and the nearby rivers and streams. It was a common man's diggings, 1850s style.

At least it was in most respects. Despite the crowding, the isolation and the inevitable disappointment, violence on the field was minimal. Placed alongside Skagway and the wild towns of the California gold rush, Dawson was a model of civic virtue. The Mounties knew how to run an orderly goldfield. After the first couple of years the population of the Klondike thinned. Of gold there was not enough, and the life was too uncomfortable. Not all the newcomers left the way they came. Some went further afield in search of new deposits, and in this way gold was found at Anvil Creek, near Nome, which at that time was a small Eskimo village, home to a handful of missionaries. It became apparent that this was a major strike, and no sooner was this confirmed than gold was found on the very beaches of Nome itself. Another rush followed, almost as big as that to the Klondike. Latecomers arrived to find the beaches pegged from one end to the other.[48] In Nome there were no Mounties, and the hastily erected town for a time was hijacked by a corrupt and menacing crew, who manipulated the administration of claims to their own advantage. But their heyday could not last. The US government now had more experience with the taming of frontiers.

In this way the last hurrah of the free miner unfolded in the wintry lands just south of the Arctic Circle. The more perceptive of the diggers who had rushed there would have realized that it was only a matter of time before technological advance introduced factory style production to the alluvial fields, just as it was doing in the hard rock mines. In fact, that was already happening. The steam-shovels of Minnesota could chew through loose gravels at a rate with which a horde of hundreds of diggers could not hope to compete. Not safe even was the gold lying submerged along the riverbeds, for even as the first diggers set out for the Klondike, in the river valleys of New Zealand bucket dredges were gouging the riverbeds for their submerged treasure. The goldfields of New Zealand had never achieved the fame of California or Victoria. The precious metal had been found along the banks of the river systems of the South Island, and many suspected that down on the riverbeds it would also be found in abundance. The depth of the rivers, however, made the bullion inaccessible with the techniques then available. The challenge thus set, the first rudimentary bucket dredge made its appearance in the early 1880s along the Clutha River near the southern end of the South Island.[49] There its successors were to remain for many years, like some species of prehistoric monster flourishing in eerie isolation. Eventually, over 200 of the 'metal mastodons' were churning the riverbeds of New Zealand.

The gold-bearing rivers of California did not get their first working dredge until much later. When at last they did, its value was quickly perceived, and a handful of dredges, the forerunners of what one historian would christen the Northern Gold Fleet, were soon put to work in the sea off the beaches of Nome. They were truly awe-inspiring creations in their time. Huge floating factories as big as ocean-going vessels, spewing diesel smoke and letting loose a noise that carried for miles. It is little wonder that, in these beasts, many finally realized the peril that faced the independent digger. All that stood between him and oblivion were the laws that, in the days of California and

Victoria, had prevented the capitalist from acquiring too large a claim. But this barrier was crumbling. Those laws now unequivocally favoured the large corporation.

The Klondike rush was three years old when the local authorities granted a huge lease, covering much of the gold-bearing ground along Bonanza Creek and the other important waterways, to an English entrepreneur, Arthur Treadgold.[50] Popular anger eventually secured the repeal of the grant, but by 1905 Treadgold and his backers had by other means managed to purchase a great swathe of claims along the creeks. His idea was to use the old hydraulicking method to hose the riverside gravels into great ponds, in which sat the ponderous dredges. Needing money to buy dredges and to build a power station with which to drive them, he teamed up with the Guggenheims to establish the Yukon Gold Company. It seems a fitting end to the Gold Rush era that the last of the great common man's diggings should finish its days being mauled by the teeth of clanking metal monsters owned by the largest mining corporation on earth.

THE DEATH OF SILVER

The gold rushes of California and Victoria had created a disturbance in the world of money. The sheer quantity of gold that flooded the globe's monetary systems raised the risk that it would eventually become so plentiful as to render it useless as the basis for a stable currency.[51] At the time, most of Europe and much of the rest of the world used a bimetallic currency, in which coins of both silver and gold were interchangeable and in accepted use. Britain and Portugal were alone in maintaining solely gold-backed currencies. The future of the money system became the subject of intense debate, and pamphlets arguing the merits of silver, the benefits of gold and the advantages of bimetallic currencies flew back and forth. As the debate wore on, gold gained the upper hand, economists and policymakers coming to perceive it as the better choice for the establishment of a sound currency. It is not wholly clear why this came about. One reason was surely that the world's most powerful economy dealt solely in gold. A second reason was probably the fact that gold's greater value per unit weight made it the preferred currency of those who dealt in frequent and large commercial transactions.

The debate continued without truly decisive action being taken in any country until, in the aftermath of the Franco-Prussian War, Bismarck decided to take his country onto gold. In the process he sold down Germany's stocks of silver.[52] Now the silver nations of Europe were faced with a problem. The German silver washing around the markets, accompanied by the torrent emanating from the Comstock, put them in real danger of serious inflation. Afraid of being left holding the silver baby, one by one the countries of Europe made the switch. Latin America followed, and by 1890 so many nations were using gold-backed currencies, and their systems were linked so closely together, that the world could be said to be working to an international gold standard.

This shift to gold was not some dry and dusty affair decided by dispassionate economists. Far from it. The issue of silver versus gold had come to symbolize the political struggle between the emerging urban industrial classes and the established rural interests. Gold, in its role as the currency of international trade, was associated with the rising urban elite of bankers and industrialists.[53] Farming communities, on the other hand, had a long history of transacting in silver, and viewed with suspicion any city-driven attempts at change. In many countries the currency question was a bitter-fought battle. Even when the matter was finally decided, the defeated defenders of silver refused to lie down. They agitated in favour of bi-metallism, and extolled its virtues in volumes of such length and tedium that their arguments eventually lost their hold on the public imagination and the topic died away. In only one influential nation did they hold their own.

Abraham Lincoln had sanctioned what all leaders sanction during times of war. He had allowed the government printing presses to churn out greenbacks in order to pay for his armies. As an inevitable result, inflation had been rampant, and well after fighting had ended the economy continued to be awash with worthless paper money. A general belief began to take hold in Washington that the currency should be brought back onto a firmer base – that is, it should be convertible to gold or silver. This raised the question: should the United States follow Britain onto a gold standard or should it return to its dual currency of gold and silver? As in Europe, it was no academic issue. Also, as in Europe, the battle-lines were clearly marked. A decision in favour of a gold standard would represent a victory for the Eastern financiers and traders who borrowed from London bankers and traded with British merchants, while the choice of a dual currency would mean victory for the farmers and for the miners of the Comstock, Coeur d'Alene and Colorado. There was a hard decision to make, and, perhaps surprisingly, the Congress was up to the task. In 1873 it declared gold to be the official currency of the United States. The cries of outrage from the West echoed through the land. It was quickly dubbed 'the crime of '73'.[54] Cowed by this storm of protest, the briefly courageous Congressmen passed the Bland–Allison Act, which guaranteed that a certain annual quantity of silver would be purchased for a fixed price and made available as coinage.

While the intensity had waxed and waned, the 'great crime' of the demonetization of silver was not forgotten. The supporters of silver could muster considerable political strength, and in fact were dominant in the American West, where there was no shortage of states in which farming and silver mining were the main industries. The senators from Montana, Nevada, Colorado, Utah and Arizona – to name only those most directly affected – were not about to let the issue die. The problem festered. The silver lobby won a notable skirmish in 1890 when the Sherman Silver Purchase Act was passed. The Sherman Act effectively increased the price the United States Treasury paid for the privilege of stockpiling silver in its already overcrowded vaults. It could not last. The pigeons came home to roost in April 1893, when it was announced to the public that the gold reserves held in the US Federal Treasury had fallen below $100 million. On hearing this news, the already nervous financial markets panicked, believing that the government could no longer guarantee its bonds. Investors rushed to convert their bonds into gold, and a rash of bank failures ensued. The failing banks, in their turn, called in their loans, and thereby created a second wave of bankruptcies among industrial and commercial concerns. Businesses failed in their thousands. Unemployment rose in the cities, and the panic settled into a world-wide business depression. In the United States, hardest hit were the farming regions. Prices for farm produce fell to unheard-of levels. Distrust and hatred of the Eastern capitalist grew within the Western towns.

In the aftermath of the crash, a determined opponent of the Sherman Act, President Grover Cleveland, blamed the depression on the United States' lack of gold reserves that, in turn, were a result, he said, of the silver purchases that the Treasury was obliged to make. In his view neither the Bland–Allison nor the Sherman Act alleviated the plight of the farmer. Pulling on every political lever at his disposal, Cleveland faced down his own party and repealed the legislation. The US government would no longer support the price of silver. Without the Treasury's backing, the silver price plunged. Coming just six months after the Wall Street crash, this was seen as a mortal blow to Western interests. The outcry was furious. Cleveland, the turncoat, had betrayed his constituents when they were most in need! Unable to stem the flood of vitriol, Cleveland was left a lame-duck President for the remainder of his term. Never mind that the Sherman Act was merely featherbedding for narrow mining interests that were quite capable of looking after themselves. The Act in the eyes of many was considered a symbol of the small man's fight against the oppressive combinations and the hated Trusts.

The popular anger swept to prominence a young Nebraska lawyer named William Jennings Bryan. Bryan was a brilliant speaker and a passionate advocate. He opposed the Trusts, the tariff and the gold standard. He supported the rights and welfare of the common man, and especially the farmers. With the repeal of the Sherman Act, he had found his national platform. As the Presidential elections approached, Bryan threw his hat into the ring, determined to fight the 'gold bugs' to the finish. At the 1896 Democratic Convention, Bryan was the last of five speakers. He gave the speech of his life, concluding with the words:

> If they dare come out in the open field and defend the gold standard as a good thing we will fight them to the uttermost. Having behind us the producing masses of this nation and the labouring interests and toilers everywhere we will answer their demand for a gold standard by saying to them: you shall not press down upon the brow of labour this crown of thorns, you shall not crucify mankind upon a cross of gold![55]

His closing phrases were accompanied by pressing his fingers to his temples followed by standing tall with his arms held straight out from his sides. The stunned audience sat silent for a moment before erupting into cheers.

It need hardly be said that Bryan carried the Convention. He then embarked on a campaign of extraordinary vigour, holding forth at small towns and railway sidings with as much energy as he would devote to a major speech in a large city. Meanwhile, the Republican nominee, William McKinley, stayed at home in Ohio, greeting visitors on his front porch. Bryan looked like he was carrying the country. The big business interests began to get worried until McKinley's political manager, a Cleveland steelmaker and shipowner named Mark Hanna, rallied the troops. He did his work so well that in the end the Republican cause was victorious, and convincingly so. The Eastern interests had defeated the West, and the menace of silver money was seen off.

The campaign was to prove Bryan's finest hour. He would lead a distinguished career, running twice more for President and serving as Secretary of State in the cabinet of Woodrow Wilson before resigning in protest at the President's willingness to involve America in World War I in Europe. His sense of moral outrage proved inexhaustible. As a lifelong devout Christian, he could not tolerate what he considered the mockery of the Bible's teachings. He was well into his seventies when he launched his last public crusade against a Tennessee teacher, John Scopes, who had committed the crime of instructing his class in the theories of Charles Darwin. In this so-called 'monkey trial', Bryan's prosecution, ably assisted by the judge, was successful. Scopes was fined $100 dollars, and the doctrine of Creationism, thus protected, continued to dim the understanding of several more generations of Tennessee schoolchildren. For Bryan, it was an inglorious end to a noble career.

McKinley's election victory meant an inglorious end to the noble career of silver. The discoveries in the Rand and Kalgoorlie helped India move to a gold-backed currency in 1893, and Russia followed three years later. Europe was solidly gold. With Bryan's loss, the privileged and powerful place of silver among metals was gone forever. The metal that had driven the mining revolutions of medieval Germany, the Spanish colonies and the Comstock had become now a minor player on the world stage. No more would it be the glittering prize, attracting the prospector and the capitalist. It lived on in one country only. China would retain its silver-based currency until the upheavals of Chiang Kai Shek and the Communist revolution pushed it back onto the paper money that the Ming had abandoned five centuries before.

GETTING THE COSTSDOWN: THE LABOUR PROBLEM

In the year when Johannesburg celebrated its twentieth anniversary, the white population of that city could afford to take some pride in their position. They were living in one of the largest cities in the southern hemisphere, and one of the most beautiful. Some, after visiting its opera house, its music halls and doing the rounds of its magnificent private homes, even pronounced it the Paris of the South – though perhaps that was the champagne talking. The economic disruptions of the Anglo-Boer conflict were well and truly in the past. In 1907 the Rand produced one-third of the world's gold, more than was produced in North, Central and South America combined.[56] Deep-level mining was now the norm, and the cyanide process universally used. Cyanidation had undergone major improvements. Instead of relying solely on hammer mills, which were now steam-powered but otherwise unchanged since Agricola's day, the mill owners had begun using revolving cylindrical mills packed with steel balls in order to grind their ores to the required fineness. Vacuum power was used to suck the gold-rich cyanide solution through a cloth filter, thereby more readily separating it from the waste ore. These tube mills and vacuum filters were now standard equipment on the Rand. Zinc dust had replaced zinc shavings.[57]

The Chamber of Mines reigned supreme in the city. It owed its success to the fact that the gold companies were now no longer competing against each other for space on the Main Reef (such was the nature of gold they never had to compete for markets). The Chamber already had at least one notable success to its name. This had been achieved in the campaign against John MacArthur and the Forrests, patent holders of the cyanide process. At first the Scottish chemists had been greeted warmly in Johannesburg, but the mood started to sour when they raised the royalty payable on their process to 10 per cent of the total gold won. Roused to action by this crippling impost, the large companies decided to challenge the validity of the patent. It was not merely a ploy on their part. In 1892 the respected *Engineering and Mining Journal* had declared: 'We are quite convinced ... that the MacArthur–Forrest patents are not valid and that the cyanide process is not now patentable.'[58] The case came to trial four years later. It was a brutal battle, drawing in witnesses from around the world. In the end, the judge found in favour of the Chamber, citing a similar American patent that pre-dated the MacArthur–Forrest process.[59] It was a big victory for the producers, for it meant that the cyanide process was now freely available to all. It was an equally big loss for the inventors. John MacArthur, having saved the Rand and then over-played his hand, died a poor man.

A decade after their courtroom victory, the cooperative efforts of the companies, organized through the Chamber, had come to focus on two purposes. First, they were concerned to establish a good working relationship with the new colonial government of Botha and Smuts. Neither side wanted the conditions repeated that led to the Anglo-Boer War. Second, they were concerned to present a united front against a common foe – their workforce.

The mineral fields of the Kimberley and the Rand were, in their labour arrangements, unlike those of the American West or Australia. Instead they resembled rather more the mines of

South America, perhaps resembling most closely of all the silver mines of the Spanish colonial era. For down in the South African mines one would not find manual labour being done by white men. It was the black tribesman who provided the muscle and the sweat. The white miner provided the 'brains' – that is, he supervised. The black man was considered the natural inferior of the European, and was treated as such.

The blacks were reluctant to work down the mines. The poor pay and inadequate food provided in the compounds in which they were compelled to live were only some of the factors that discouraged them. The great majority of the labourers were 'hammer boys'. Their task was the backbreaking work of hammering holes into the rock so that explosive could be packed in. Death through rockfall or disease was common. In addition, the time away from home, the toil in the darkness and the strange nature of the tasks held little appeal for the majority. The warrior who ventured underground was considered to be embarking on a perilous adventure.

Thus, as the mines expanded, the Randlords found themselves perpetually short of labour. They tried importing Chinese coolies, but the political outcry against cheap Asian labour was so violent that the Conservative government in Britain lost office and the plan was abandoned.[60] Denied the option of importing cheap labour, there was no option but to improve the working conditions and pay of the black man in order to attract him underground. Compensating cost reductions had, then, to come from two areas – higher productivity and a reduction in the cost of the white miners. On the productivity front the development of the lightweight air-powered jackhammer gave a great boost to the daily output of the hammer boys.[61] As for the white worker, he was prey awaiting a determined hunter.

Sensing this, the white workers began to take what defensive action they could. The membership of the Transvaal Miners' Association, representing white workers, began to swell, and when the companies made their move the miners were prepared. The trigger was a minor local dispute involving five men and Saturday working hours. It is a measure of the tension on the goldfields that such an incident could spread into a general strike. In July 1913 the flames of the dispute finally flared to engulf the Rand.[62] That night mob rule descended on Johannesburg. Battles with police and then mounted British troops ended in defeat for the miners, but only hardened their resolve. The following day, as the situation threatened to deteriorate into a bloodbath, Generals Botha and Smuts, the two most influential men in the government of the Union of South Africa, met the strike leaders at an inner city hotel. Outside, a mob raged and heckled. While the pair agreed to the strikers' terms, it would prove to be a short-lived victory for the miners. When they downed tools the next year it became clear that the government's patience had been pushed beyond its limit. Martial law was declared. Faced with implacable opposition from their war-hardened leaders, the ardour cooled in the hearts of most of the strikers. A few of them holed up in the Trades Hall. In answer Smuts wheeled into position a heavy field-gun. The men, he declared, must surrender within five minutes, or the Trades Hall would be reduced to splinters. So ended – peacefully – the 1914 Rand general strike.

With that came a truce on the goldfields, as Randlord and worker banded together to fight the Great War. For the companies the few gains of 1914 were whittled back during these years. After the War, working together through the Chamber of Mines, they renewed their cost offensive with fresh resolve. Protecting their position, the white workers had one major factor in their favour. This was the Colour Bar, a body of legislation that nominated a wide range of jobs as being the sole preserve of the white man. The companies began by challenging the Bar, assigning to black workers more and more tasks that had been traditionally undertaken only by a white man. The next step was to confront the union with a demand for the removal of the Colour Bar. It was thrown out. Many considered this a deliberate provocation. The companies, they perceived, had now embarked on more than just an industrial action.

The attempted removal of the Colour Bar struck at the very foundation of the white man's position in this tiered society. Union agitators cried that white supremacy in South Africa was under threat. Nationalist Boers organized commandos to defend their rights, just as they had defended their land against the British. Sensing a dangerous conflict brewing, Smuts, now Prime Minister, wanted compromise, but the President of the Chamber of Mines, Lionel Phillips, was reported to have replied: 'Who should be running this place?' In January 1922, coal miners walked off the job, followed by the whites working the Rand mines. The trouble escalated until the union called a general strike. In the days that followed, about 10,000 strikers occupied the shaft heads and other strategic points along the Reef. The newspapers called it the Rand Revolt, the term capturing the political nature of the strike as well as the strikers' near-military organization. Amid the hysteria the companies convinced themselves that the Communists had taken over the field. Perhaps in a vague way the workers, too, thought of themselves as allied to the international Communist movement, a possibility that would help explain their oddly garbled slogan, 'Workers of the world fight and unite for a White South Africa'.[63]

For four days gun-battles shook the city. Military aeroplanes conducted strafing missions against their enemy. Of police and soldiers, the death toll had climbed to 72 before the last pockets of resistance were crushed. Of strikers, or 'revolutionaries' as the government preferred, the toll (surely underestimated) was 39.[64] Another 42, mostly blacks, died in the crossfire. In the aftermath the mining companies pushed their advantage to the limit. White wages plummetted, thousands were laid off and many mines totally reorganized their operations. Yet for the unionists and the Boers, the strike was a victory of a different kind. The Colour Bar legislation held firm, kept in place by a farmer-dominated Parliament. Eventually it became the symbol and rallying-cry of the white workers. Smuts, the war hero and visionary statesman, had destroyed himself. White voters could no longer trust him. In the next general election they turned him out of office, choosing instead a Nationalist/Labour coalition led by General Hertzog. Hertzog's government set about strengthening the economic position of whites at the expense of blacks, in the process laying the foundations for what would become the system of apartheid.

The savagery of the Rand Revolt does not stand alone in the

history of the period. In 1914, during a strike at the Rockefeller-owned Ludlow coalfield in Colorado, paid thugs and state militia poured petrol on the tents of striking workers and set them alight. Nineteen women and children were burnt to death, and miners were shot dead as they fled the blazing camp.[65] Three years later, in the Arizona copper town of Bisbee, more than 2,000 striking workers were rousted out of their beds at rifle-point and taken to the railroad station, where, under the gaze of a pair of machine-guns, they were given a choice: return to work or leave the state. That same day, over 1,000 men were herded onto boxcars and railed across the state line.[66] Bitter and violent confrontations such as these were for a time a feature of mining towns throughout the world. Mostly they were about the struggle to contain costs in an over-supplied industry. To some it was a sign of the times. The unrestrained capitalism of the last few decades had run headlong into the new doctrine of socialism. Perhaps it was this ideological divide that can explain the extreme polarization over issues that, on the face of it, seemed amenable to more peaceful resolution. The fear of socialism was a factor during the Rand Revolt. In the case of Bisbee, the strikers were members or sympathizers of the International Workers of the World – a militant union that took seriously its socialist principles.

Perhaps, too, the tensions were raised above normal limits by the fear of phthisis, the deadly lung disease that was killing mine workers around the world as they inhaled the fine dust generated by the action of the jackhammer drills. Of the twelve leaders of the Rand strike of 1913, phthisis would kill seven. Only in Australia did deadly force remain largely absent from industrial confrontation. There the political strength of the large companies went nowhere near that of their counterparts in the United States, Latin America and South Africa. The dispute that for eighteen months shut down the Broken Hill field unfolded without a single fatality due to strike-related violence. Justice Edmunds finally ended the stand-off by granting the men a 35-hour week and generous pay rises.[67] In time, as socialist ideals gained wider acceptance and were absorbed into the social and legal framework of many nations, much of the rest of the mining world would follow the spirit of Broken Hill.

Amid the labour battles, the excitement of discovery retained its power to fire the imagination. Exploration on the Rand was dominated by the effort to solve a mystery. East of the twenty-mile stretch of shafts and mills that feasted on the Main Reef, the ribbon of banket suddenly vanished. Did it just peter out, or had some ancient cataclysm dislocated it? If so, where had it gone? Small fortunes had been lost in the search, and over time observers in the older and prosperous companies had drawn their own conclusions: exploration did not pay. They left the search for new gold to other, lesser companies. Back in the Kimberley, too, stagnation had set in. The diamond trade had become the establishment business of South Africa. Rhodes and all the others had been right. Amalgamation had done wonders for the industry. Fourteen years after the epic duel it had become fatly prosperous. The wealth and respectability had in turn sapped its vitality. Rhodes himself had pronounced the business boring as he pocketed its profits and used them in order to further his empire-building schemes.

So assured of their God-given right to dominion over diamonds had the men of De Beers become that news of a rich discovery near Pretoria was waived away. Obviously, sniffed chairman Francis Oats, the ground had been salted to fool the gullible.[68] He, no doubt, drew reassurance from the fact that the alleged discoverer, Thomas Cullinan, was a mere plumber. The men of De Beers were in for a shock. In only six years Cullinan's Premier mine was selling more diamonds than they were. Adding insult to injury, one evening in 1905 the Cullinan diamond was found, the largest ever discovered. Then diamonds were discovered in abundance in the sand dunes of the German colony of South-West Africa. Again Oats and his colleagues sat on their hands. Now there were three separate producers competing for a slice of the diamond market.

Looking down, or up, on all these happenings, Cecil Rhodes would not have been pleased. His successors were bland men, complacent and cautious. South African mining, he would have seen, needed a fresh a man with the imagination to create a bold vision and the strength of will to see it to reality. Would he have guessed that already present in the Kimberley was the man for that task? This was Ernest Oppenheimer, the quietly spoken diamond trader who was making a name for himself in Kimberley. Oppenheimer was born in the town of Hesse into a German-Jewish family that later emigrated to England, settling in London. There, aged sixteen, Ernest entered the diamond business of Anton Dunkelsbuehler. Dunkelsbuehlers & Company was a member of the buying cartel that alone was permitted to purchase the De Beers production. Anton was impressed by the quiet determination of his new recruit. Having amassed sufficient experience in the London trade, in 1902 Oppenheimer was despatched to the Kimberley as representative of Dunkelsbuehlers. It was the year in which Cecil Rhodes died.

The Kimberley stint was the usual tour of duty for a capable scion of the close-knit diamond-dealing community. No doubt after some years Oppenheimer would be expected to return to the comfortable setting of the London markets. But he found South Africa to his liking. As the years passed, he gained the respect of the De Beers crowd and earned the complete confidence of Dunkelsbuehler. He secured himself a place in Kimberley society. He was elected mayor of the diamond town. Then came World War I. The focus of worker discontent turned towards the German contingent in both Kimberley and Johannesburg. The North Atlantic sinking of the *Lusitania*, a passenger ship, by a German U-boat, provided the excuse to unleash their bottled-up passions. British citizenship was no protection from the wrath of the crowd. In Kimberley, the Oppenheimer home was attacked by an angry mob. In Johannesburg, the Randlord George Albu hastily scribbled a will as rioters besieged his mansion. To Oppenheimer, mayor of Kimberley, the affair was an intolerable insult. In disgust he sailed for Great Britain.

The years in South Africa had not been wasted. Oppenheimer had observed at close hand the situation in Kimberley and Johannesburg, and knew that there lay opportunity for a bold and vigorous entrepreneur. Apart from their complacency, the established mining houses had a major weakness. They were all London-based. Their boards met and decided the issues of the day from the point of view of the City's financial community. Few of the company directors had been to South Africa in years. The absentee Randlord did not sit well with the new leadership of the Union of South Africa. Thus, when Oppenheimer began discussing his plans to create a new South African-based mining house, he received encouragement and useful introductions from government ministers. All the while he formulated his strategy. The proposed mining house would aggressively seek and develop properties in the Far East Rand. The most crucial introduction came through his trusted friend the American mining engineer Charles Honnold. Honnold arranged a meeting with Herbert Hoover, who was now a doyen of international mining and increasingly a political figure of influence. Via Hoover, Oppenheimer secured the financial backing of J. P. Morgan and of William Boyce Thompson, founder of Newmont Mining.

Now equipped with funds to underwrite his plans, there remained the task of selecting a name for the new company. Its working name was Far East (Rand) Trust. But to Oppenheimer, this was too limited. He wanted something that captured his vision and would appeal to financiers in the important capitals of the world. Thus in 1917 in Johannesburg was launched the Anglo American Corporation of South Africa.[69] With the float of Anglo American, Oppenheimer moved on the Far East Rand. Alliances with existing companies allowed him to participate in the acquisition of leases and the development of several significant gold mines. Other companies – those without a strong, deep level base on the Main Reef – were also trying their luck in the area. Solly Joel was one of the most prominent, as he guided Barney Barnato's JCI back to its former greatness. But not even he could match the drive, cunning and dealmaking ability of Oppenheimer. Not having to refer to his board in distant London, Oppenheimer could move with unusual speed. In short order Anglo American had secured a stake in the burgeoning Far East Rand and was a force on the goldfields. It was time for Oppenheimer to turn his attention to his first love.

The diamond fields of South-West Africa had fallen into British hands at the end of World War I. Negotiations surrounding the carve-up of the former German territories were still underway when some directors from Anglo American arrived in Versailles to speak with General Botha. Would the General object to a bid for the diamond properties? With Botha's blessing, Oppenheimer bought out the German owners.[70] The diamond establishment felt outrage. They had been betrayed by one of their own. Some never forgave him. Oppenheimer paid little heed to the outcry and instead remained focused on his burning ambition. Where new sources of diamonds were found, there too would be the quietly forceful chairman of Anglo American. In Portuguese West Africa, in Sierra Leone and in the Belgian Congo, there was Ernest Oppenheimer, closing deals and broadening his power base in the industry. In the process, he cultivated Solly Joel.

By 1925 Oppenheimer was strong enough to make a successful bid for control of the London-based diamond-selling syndicate. There now remained only the task of seizing control of the De Beers diamond production. Having long since lost their monopoly, De Beers was a fat goose ready for plucking. Oppenheimer made his move by giving Solly Joel a choice: support me as De Beers chairman or face a ruinous diamond price war. That year Oppenheimer was elected chairman of De Beers. He now sat at the head of the South African mining establishment. Anglo American would go on to establish a dominant position in the world's platinum market, and it would take a leading role in the international consortiums that financed the copper developments of the Congo. De Beers would maintain its world-wide diamond cartel for two more generations. As a *coup de grâce*, Oppenheimer's company would eventually become the largest of the Rand gold-mining houses.

The international gold standard collapsed during World War I and, despite attempts at reconstruction, never reappeared. The stresses of the Great Depression and World War II finished it for good.[71] It had had an effective life of perhaps 40 years. With the European economies having moved onto a paper-based currency, it was left to the United States alone to maintain a gold-backed currency. During the Depression, President Roosevelt set the greenback conversion rate at $35 to the ounce. And there the price stayed, year after year, until the turmoil of the first oil crisis forced a different President,

Richard Nixon, finally to abandon the scheme. Throughout this period, the banket reefs became by far the greatest gold-producing region on earth. They were low-cost and high-volume. With the Federal Reserve refusing to budge on its price, there was little incentive for anyone else to develop new mines. In 1970 South Africa's share of world gold production came close to four-fifths.[72] The latter-day Randlords had the field to themselves.

SIX

Mass Production

THE HEART OF EUROPE

Remember the Habsburgs? In 1914 the family still occupied a space at the centre of European power. To be sure, the seat of their power, the Austro-Hungarian empire, had declined of late. Some even called it senile, and likened it to the tottering ruin that was the Ottoman empire. The fundamental difficulty for both was that they had failed to adapt to the currents of nationalism and democracy that were flowing through Europe. The peoples within their borders wanted their freedom, while their monarchs continued to cling to older notions of imperial splendour. Economic torpor accompanied social unrest. The contrast with the German Empire could not have been greater. Here was a people united, vigorous and enterprising. German universities were the best in the world. Her chemists and physicists led the way in discovering new scientific principles and new industrial techniques and processes. Heavy manufacturing had grown so rapidly that the country could now fairly claim the title of the leading industrial power in Europe. Electrical equipment was one specialty, chemicals were another. Precision tools and instruments of all kinds were yet another. Such industry had need of all manner of metals.

Of copper, some was mined within German borders in the old faithful fields of the Harz and Saxony. In Silesia deposits of lead and zinc were worked. Bauxite was mined in France and smelted to aluminium in Switzerland, all under the auspices of the new German electrical giant AEG. Most of the Reich's non-ferrous metals had, however, to be shipped in from across the ocean. In this business of importing ores and metals, one company had become dominant. It was founded by Wilhelm Merton, a principal in the venerable Hamburg-based metal trading firm of Merton & Cohen.[1] Merton was from the mould of the new breed of German businessmen. He perceived the changes in the world of industry, and realized that the old metals-trading firms such as his own were faced with a choice. They could continue in their comfortable traditional methods and be crushed between the German factories and the great metals conglomerates growing up across the Atlantic, or they could confront and master the developments on the world stage. Wilhelm Merton intended to do the latter. He formed Metallgesellschaft (literally, the Metal Company) and not long after set up a sister company in London. In Hamburg he built the first electrolytic copper refinery in mainland Europe. In the United States he established the American Metal Company, after which, in rapid succession, came investments in Mexican mines, Western smelters and New York refineries. In 1912, in the Belgian port of Hoboken, Merton opened yet another refinery to receive the copper just beginning to arrive from the Belgian Congo. His tentacles stretched to Australia. There his Australian Metal Company purchased the zinc concentrates of Broken Hill and shipped them to his smelters in western Europe.

At the heart of German industry were its blast furnaces. Day and night these manufactories pumped their molten iron into the arteries and capillaries of the industrial complex. The greatest concentration of steel works could be found in the Ruhr, a coal-rich region near the border with Holland and Belgium. Let us take a moment to appreciate the geography within which the Ruhr is situated. The Rhine river has its source in the Swiss Alps. Its first 100 miles or so separates the provinces of Alsace and Baden. Alsace and its neighbouring province of Lorraine have a special place in European history. This is because they form the border between France and Germany. Nestled up against Lorraine is the coal-rich tranche of territory known as the Saar. The Rhine then winds its way through Germany. The ancient city of Cologne sits astride it. Just north of Cologne it collects the waters of the Ruhr and Emscher rivers before crossing the border into Holland and emptying into the sea at Rotterdam. Not far to the south-west of Cologne can be found the old brass-making cities of Aachen and Liège. The region

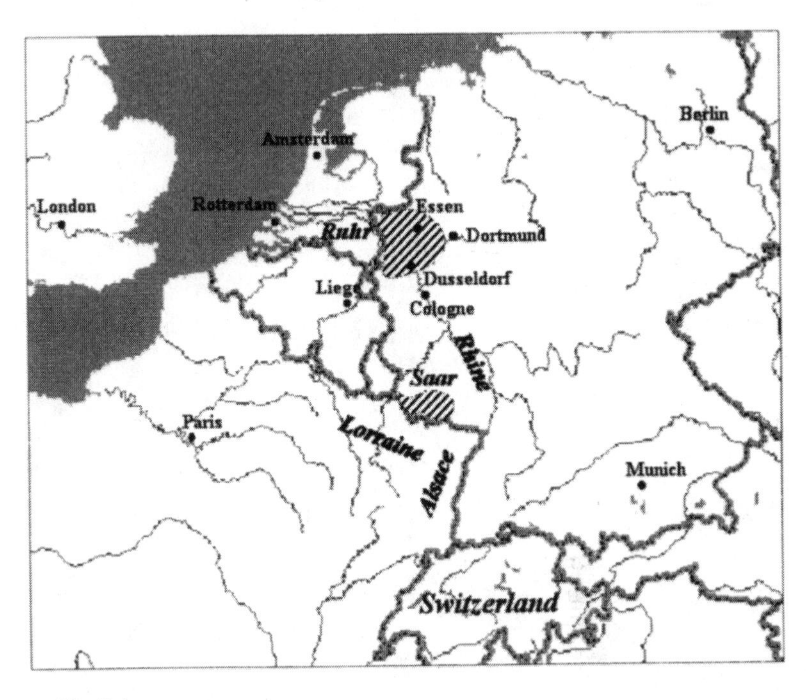

10 The Ruhr, 1910

defined by Rotterdam, Cologne and Liège has long been one of the more densely populated parts of Europe. Just outside this triangle, to the north-east, is the Ruhr.

The importance of the Ruhr in the German economy had its origin in the rapid industrialization that followed the Zollverein. This was an agreement struck in 1834 among the many semi-autonomous German states. It did away with the system of tariffs that had created what amounted to economic gridlock throughout the German lands. For centuries the potential of the people and the land had lain dormant; now, under determined Prussian leadership, that potential began to express itself once more. A network of railways was built, and industrialization proceeded apace. Coal was needed. At the time the chief source of German coal was Silesia, far away in what is today Poland. Now it was realized that the Ruhr and the Saar contained plenty of readily accessible, good-quality coal.[2] Being far closer to the main markets, soon the two regions were the scene of vigorous mining expansion. Iron works were opened up. The boldest and most dynamic of this new generation of iron masters was Alfred Krupp whose works could be found in the city of Essen. Son of the firm's founder, he had taken charge at the age of fourteen. Like Andrew Carnegie, he had stayed aloof from the clubby iron industry of the time and had fixed his purpose on making low-cost products for the country's expanding railway system. He built the region's first Bessemer steel mill, and turned his hand to the production of rails and cannon. At first, out of a mixture of snobbery and conservatism, the Prussian army had shunned his armaments. But war has a habit of sweeping aside self-indulgence. When German armies invaded France in 1870, the generals soon found that a Krupp cannon was a faithful friend. It played a crucial role in putting to flight Napoleon III's ill-prepared and over-confident soldiers.

In the aftermath of the French surrender, the victorious Prussian army insisted, contrary to Bismarck's urgings, on the annexation of Alsace and about half of Lorraine. One might be excused at this distance in time for suspecting that the annexation of Lorraine was motivated by a desire to secure the iron fields for which the province later became famous. This was not the case. For one thing, the extent of the iron reserves was not then understood, and, second, what *was* known about them was that their quality was worse than those commonly found in the Ruhr. And the iron ore of the Ruhr was poor-quality enough. So much so that the German steelmakers of the time routinely used Spanish ores to supplement their own. The trouble with the iron ore of the region was that it was full of phosphorus, the very element that had come so close to destroying the Bessemer process. It presented a real obstacle in the quest to produce high-quality steel at low cost. Far from wanting to acquire more of it, the German steelmakers were already engaged in a determined search to find a way to use the reserves they already possessed. Krupp's metallurgists were the first to find a solution, and at nearly the same time, on the other side of the English Channel, two cousins also solved the puzzle.

Sidney Gilchrist Thomas was a 20-year-old clerk attending chemistry lectures in his spare time when the phosphorous question came to his attention.[3] He had scant formal training, little money and no obvious reason to concern himself with the problem. Which makes it all the more curious that from 1870 to the day in 1879 when he delivered his paper to the Iron and Steel Institute, Thomas pursued the problem with singular intensity, assisted at times by his metallurgist cousin Percy Gilchrist. His eventual solution was simple enough: by lining the blast furnace with limestone bricks, the phosphorus would be soaked up out of the molten iron. Cheaper and more practical than the Krupp method, the Thomas–Gilchrist process unlocked the potential that lay latent in the Ruhr. The story goes that, on hearing of the breakthrough, two Ruhr-based firms despatched representatives with instructions to cross the Channel in all haste and purchase the rights to the process. The race, it is said, was won by the man who did not stop to sleep.

With the phosphorus problem solved, the steelmakers regained their purpose and the next decades were a time of intensive modernization throughout the Ruhr. Essen, Duisburg, Dortmund and Bochum all became leading steel cities, their skylines a forest of smokestacks. Larger and more efficient plants were built. In pursuit of cost advantage, the steelmakers purchased coal mines and integrated forward into the fabrication of finished goods. When a large and high-grade source of iron ore was discovered in Sweden's north, a canal was built from Dortmund to the coast to minimize the cost of delivering it to the mills.[4] German industry, alone in Europe, was following a path similar to that being forged in the United States. Price competition became brutal. On the back of the modernized mills, German steel production passed that of Great Britain around the turn of the century, and by 1910 the country was easily the leading steelmaker in Europe. The Krupp conglomerate, with 70,000 employees, remained the largest. It had by now become the establishment steelmaker, and enjoyed the status of favoured supplier of armaments to the Kaiser's army and navy. For them, business was especially good. An emboldened armed forces needed guns, cannon, ammunition, vehicles – the list went on.

The French neither forgot nor forgave the loss of Alsace and Lorraine, but far from being in a position to reclaim their territory, they looked on with apprehension at the growing military might of their neighbour to the north and east. Other nations, too, regarded the Germans as arrogant bullies who might one day push Europe into war. At times tensions rose to flashpoint. Yet, by 1914, many were thinking that the worst was over. Someone published a book that proved that international commerce was now so interlinked that war was, in fact, impossible. Many nodded their heads in agreement. Then came the news of the assassination of Archduke Franz Ferdinand, heir to the Habsburg throne. The ill-considered imperialism of the Habsburgs had finally caught up with them. The Archduke and his wife were shot dead in the coastal city of Sarajevo in June 1914 by a Bosnian nationalist unhappy about the annexation of his country into the Austro-Hungarian empire. In August of that year a bewildered Europe found itself enmeshed in a general war, the first ever between industrial powers. It was widely prophesied that, without swift victory, Germany must necessarily succumb. After all, the German and Austro-Hungarian lands were without mineral wealth. From where would they source the copper and zinc that were essential to the production of shell casings and other ordnance? And what of nickel and chrome, the metals that were indispensable to the manufacture of hardened steels? A Germany at war must also make do without manganese for standard steels, and neither had it access to supplies of good-quality bauxite. In Britain, the government pushed Henry R. Merton & Company into liquidation. Its sister company's zinc concentrates were left stockpiled at Broken Hill. A British blockade prevented Chilean ships laden with nitrates – an essential ingredient in explosives – from reaching Baltic ports.

Despite all these difficulties, the German war effort was never seriously hindered for lack of metals. The mines of Serbia were plundered for all they were worth. The old copper-fields of Agricola's day were dragged back to life, and a scrap recycling programme, a model of methodical efficiency, was instituted to gather everything from discarded copper wiring to household utensils.[5] Soldiers picked exploded shell casings from the cratered battlefields of Belgium and France. In May 1916, and then again in November, a U-boat, the *Deutschland*, docked in New York and departed with its hold full of nickel, tin and rubber.[6] French prisoners of war worked the mines of Lorraine. The deposits of Polish Silesia continued to pour forth their zinc in such abundance that it was used in place of copper in many applications. Indeed, the German scientists and engineers developed a genius for substitution. Iron and steel were adapted to all manner of novel purposes. Aluminium, extracted from low-grade clays, was the star of the show. Once shunned by metal workers, this new material now found its way into all sorts of applications that hitherto had been the preserve of steel and copper. And a process was developed for manufacturing nitrates from nitrogen captured out of the air. For all the intensity of their blockade, not even the British could prevent fresh air from falling into the hands of the enemy.

Instead of a shortage of metals, it would be a shortage of food that, after four years of brutal trench warfare, finally pushed the Kaiser and his generals into surrender. In the aftermath, the Ottoman and Habsburg empires sank at last beneath the waves of nationalism. Further east, the Bolsheviks succeeded in bringing revolution to Russia. In western Europe it was the Franco-Prussian War all over again, but this time the French were the victors. A vengeful peace was imposed. The stolen lands of Alsace and Lorraine were reclaimed, the Saar placed under League of Nations administration, and a huge reparations bill was thrust on the Germans. Amid a clearly unstable international scene – in 1919 the leading French general, Joffre, had pronounced the peace 'a ceasefire for twenty years' – the world got back to business.

THE CARTEL ERA

For the first time in history, industrialized nations had locked together to fight a 'total war'. Great strains had been placed on the metallurgical industries of the world. Global production of copper, zinc and steel had increased between one-third and one-half during the five years of conflict. The governments of the belligerent states had not sat back and hoped that the free market would deliver such rapid growth. Men were appointed whose task it was to unify and greatly expand the national factory system.[7] It was as though the productive forces unleashed by the second industrial revolution had come together in one last mighty heave. The greatest beneficiary was the United States. Its dominance of global metal production was already overwhelming before August 1914. Now, in 1918, American steel production stood at 45 million tonnes, up from 31 million in the last year of peace. And 1913 had been a record year. In addition to its global leadership in steel, at war's end the United States smelted and refined three-fifths of the world's copper and zinc, three-quarters of its aluminium and half of its lead. It ran a close second to Mexico in the production of silver, accounted for one-fifth of global gold output, and for good measure produced three-fifths of the planet's supply of petroleum, two-fifths of its coal and two-thirds of its sulphur.[8] During the course of the struggle, the mines of the American West, Chile and Mexico had been ruthlessly exploited. Smelting capacity had also grown rapidly, though not in proportion to output, because the war also saw heroic improvements in productive efficiency.

Now with the guns silent and the military machine idle, demand for metals slumped. Productive capacity, of course, remained in place, and for the first time in living memory some public sympathy was aroused for the notion of peacetime combination in business. Everyone knew the dangers of this, and the man to whom fell the task of achieving the right balance of cooperation and competition was Herbert Hoover, the Secretary of Commerce in the cabinets of Warren Harding and Calvin Coolidge. In 1922 he outlined his views on the matter:

> There has been a profound growth of understanding of the need and possibilities of co-operative action in business that is in the interest of public welfare. Some parts of these co-operative efforts are inhibited by law today [and] many are stifled out of fear or shackled from uncertainty of the law ... Where the objectives of co-operation are to eliminate waste in production and distribution, to increase education as to better methods of business ... to negotiate collectively with highly organized groups of labour, to prevent unemployment; then these activities are working in the public interest.[9]

Hoover was 48 years old. He had made the transition from mining to politics during World War I when he took on the task of delivering wartime food relief to Belgium, and then was given a similar job in his own country. He was later to take charge of food-relief efforts in post-war Europe and in post-Revolutionary Russia. During his busy mining career he had found the time to complete, in collaboration with his wife, the painstaking task of creating the first modern translation of *De Re Metallica*. He had also written *The Principles of Mining*, a text-book that would help educate three generations of mining school undergraduates. He was no angel. As a businessman he had taken an aggressive approach to both competitors and workers, and his past included a rather questionable takeover of the Kaiping coal concession near Shanghai, which had made him the subject of a British Parliamentary enquiry. He would prove to be a popular and principled politician, eventually being elected President of the United States in 1928. It was an ill-starred Presidency. Only nine months into his tenure came the Great Crash on Wall Street, and he would spend the rest of his term struggling without success to lead his country out of the deepest economic depression in its history. A brilliant and forceful humanitarian, Hoover's most enduring legacy would be the Hoovervilles, shanty towns erected in Washington, DC, by the protesting unemployed.

The business programmes pursued by Hoover were a moderating reaction to the trend of the pre-war decade. That was the period during which the hard work had been done of breaking the stranglehold of big business on the American national welfare. The pioneer in that struggle had been President Theodore Roosevelt. On succeeding the pro-business William McKinley, following his assassination in 1901, Roosevelt soon showed himself to be a man of different stripe to his patron. He set out to break the power of the Trusts, targeting first one of J. P. Morgan's more blatant monopolies, a combination of the three major railroad systems in the country's north. Others followed. His chosen successor, William Taft, accelerated the campaign. pressing anti-trust suits far and wide.[10] The biggest casualty of the period was Standard Oil. That monolith was carved into seven separate entities after emerging the loser from a Taft administration anti-Trust suit. The metals Trusts had survived. In 1911 Taft took aim at US Steel. The next year it was the turn of the Aluminium Company of America (Alcoa), previously known as the Pittsburgh Reduction Company. Both companies emerged intact from their ordeal, although Alcoa's survival was said by some to be owed to the connections of one of its board members, Andrew Mellon, the banker who later would work alongside Hoover as Secretary of the Treasury. Somehow or another the Guggenheims were not even brought to the stand, although Roosevelt had taken an especial dislike to their business methods.

Hoover's stance allowed the industrial giants to breathe more easily. US Steel, the biggest company in the nation, set the style. Having distinguished himself through US Steel's success in expanding wartime production, in the post-war years Chairman Elbert Gary did his best to project a statesman-like image to competitor and public alike. Even before the war he had proceeded down this path, hosting dinners to which all industry leaders were invited and, when the legality of such gatherings was challenged, forming the American Iron and Steel Institute to promote what he called friendly competition. His stewardship of the steel industry was so benevolent that during the long drawn-out proceedings of the anti-Trust case, his competitors gave evidence in his favour. (This case, initiated during the Taft administration, grew out of US Steel's takeover in 1907 of the Tennessee Iron & Coal Company.) His toughest opposition came from within his own camp, especially from Carnegie's men, who could not reconcile themselves to the gradual decline in market share that was the result of his policy. Still, Gary's style suited the time. Asarco, Anaconda, Alcoa and International Nickel followed his lead and adopted similarly careful strategies. The result was an environment where understandings and published price lists ensured a reasonable profit for all. When their production exceeded domestic demand, the American metals firms ventured onto the international market.

Here they encountered a different world. It was one in which cartels formed, broke up and formed again in an unceasing effort to bring order to the chaos. The Europeans, politically fragmented and borne along on a tide of nationalism, had not managed to tame the industrial beast. Economies of scale and a rush for national self-sufficiency, particularly in war *matériel* such as metals, had resulted in a chronic excess of capacity. Complicating matters, when hostilities ceased, Europe had found itself tossed about on a sea of floating currencies. One month French steel might be cheap on the wider European market, the next it would be expensive. When, on top of all this, they were faced with the prospect of competing against the Americans, it is little wonder that European smelter men sought refuge in the shelter of a cartel. Market share agreements began to flower immediately after the war, and in the decade that followed international cartels were formed in tin, lead, zinc, copper, nickel, steel and aluminium to name only the most prominent. Legislation banned the Americans from participating in these syndicates, but this did not mean that they were forced to step forth from their shores totally unprotected. In 1918 an understanding Congress had passed the Webb-Pomerene Act, which made it legal for American companies to band together in order to compete more effectively in the international marketplace. Inevitably, these Webb-Pomerene associations found ways of reaching mutually acceptable accommodations with the European cartels. So it was that the world of metals evolved to the point where it was held together by a grid of market share agreements, pricing arrangements and interlocking ownership and directorships. The agreements were unstable. The temptation to cheat was overwhelming, and the more aggressive companies complained constantly of being allotted unjustly low shares of the market. But when the agreements fractured, it usually took only a few years before most companies saw the sense in returning to the table.

The case of zinc is typical.[11] It had taken several years for the zinc producers of Europe to recover from their wartime dislocation. The Belgian and French production capacity located in the main battlefields had, it need hardly be said, suffered severe damage. As they reinstated their smelting works, these producers sought anew for ores with which to feed them. Their Australian source was gone – monopolized by the British, who had, during the emergency of war, rather hastily put together a rescue package that guaranteed the purchase of all Broken Hill zinc concentrates until 1930. Undeterred, the Belgians ventured into Mexico, the United States, Bolivia and Canada, signing up tonnages wherever they could find them. Meanwhile, an Anglo-Australian zinc bloc had come into being based on the Broken Hill concentrates and the Bawdwin mine of Burma, into which Herbert Hoover had poured years of work and millions of pounds. Canada entered the act by creating its own industry based on the extensive underground deposits of the Sullivan mine and the development of the electrolytic zinc process, a technique that had at last been perfected at Trail in the western Canadian province of British Columbia. The Silesian fields, now lying within the borders of the newly created republic of Poland, continued to supply the German factories. When, in 1925, an American Zinc Export Association was formed under the Webb-Pomerene Act, it seemed to producers everywhere else that something had to be done.

The first attempt failed because the British could not accept the Belgians as leaders of the cartel. As the market situation steadily worsened, however, the companies were forced back to the table. In 1928 a discussion group was formed. This led, six months later, to an agreement to curtail the output of each producer. It was a promising beginning. Yet the seeds of its destruction were already sown. Silesian producers were dissatisfied. Neither was allowance made for expansion of Canadian and Australian production. And then there was an unexpected revival in demand. Unable to cope with the divergent pressures, the cartel dissolved after less than a year. The catastrophically low Depression prices catalysed a renewal of the agreement in July 1930. This time the electrolytic refineries, with their lower cost structures, cheated by not reducing output. The cartel dissolved again. Not to be put off, the Belgians tried once more, brokering a five-year accord that was signed in 1931. The agreement was revised in 1932, in 1933 and again in 1934. It would stagger on in this fashion until its final collapse following the declaration of war in Europe. On each occasion, real success in limiting production and pushing up the price was hampered by dissent, bending of the rules, outright cheating and eventual dissolution. The remarkable aspect was the doggedness with which the leading groups overcame their disappointment and frustration and took up again the challenge of forging a fresh agreement.

One of the greatest threats to a cartel is the entry of an aggressive new competitor with the ability to undercut on price and the size to make its presence felt in the market. In such a case the cartel members have a choice: they can make room for the newcomer, or prepare themselves for a fight to the finish. This was the situation faced by the American producers when copper was found in northern Rhodesia, just to the south of the border with the Belgian Congo.[12] Union Minière was prospering. It had constructed a town, Elisabethville, and all the infrastructure necessary to mine, smelt and rail a growing quantity of copper. It had grown to become one of the largest producers in the world, and a committed member of the global cartel. The company was confined to activities within the Belgian Congo. It was known that the copper-field it was working extended across the border into Rhodesia. Various parties – among them Chester Beatty, John Hays Hammond's protégé, and Edmund Davis, the so-called father of Rhodesian mining – held leases in the area, but the ore was considered too poor in grade to make development worthwhile. That was until a drilling programme supported by Ernest Oppenheimer showed a treasure trove lying at depth. Between 1926 and 1928 in northern Rhodesia, five separate major discoveries were made.

Asarco tried to push into the scene.[13] That it failed in this effort may be put down to the nationalism of the day. Oppenheimer convinced the British Prime Minister that American involvement in so large an undertaking was contrary to British interests, and the Prime Minister duly called on his friend Sir Auckland Geddes, chairman of Rio Tinto, to step in with capital to assist Oppenheimer. It is testament to the continuing financial strength and dealmaking acumen of the American firm that it would take these two companies along with Rothschild money and the staunch support of the British government to repel the interlopers. The spurning of Asarco might have led to a battle royal along the lines of the old Anaconda–Lakes Copper struggles, but this did not match the temper of the times. Instead, when the Rhodesian copper was ready for market, the phalanx of the copper cartel parted, and when it closed once more, there stood the newcomers shoulder to shoulder with Asarco, Union Minière, Anaconda, Rio Tinto and Kennecott. The affair said much about the way in which global mining was developing. The Trust movement and the cartels had created powerful and inter-linked groupings that sat astride mineral activity throughout the world. In aluminium, too, a powerful cartel spanned the globe, controlled by only a handful of companies.

THE RISE OF ALUMINIUM

The election of 1896 had upset the natural order of things. For millennia the world of metals had been dominated by the quartet of gold, silver, copper and iron. Now silver was gone. Not completely, of course. It remained in demand as jewellery, especially in India, where it doubled as a store of value for Hindu wives who were not entitled to assume any of their husband's wealth should he die. Eventually it would also be employed as a coating agent for photographic film. As for the remaining trio of gold, copper and iron, they were soon joined by another. Its appearance was silvery and its original uses ornamental, but aluminium would establish its place as an industrial metal. It was a meteoric rise to prominence.

Many had grasped the inherent advantages of aluminium in the years following the invention of the Hall–Héroult process. Here was a lightweight metal that resisted corrosion, was easy to shape and could conduct electricity nearly as well as copper. Despite all this, and the strenuous boosting of enthusiasts, in its first ten years aluminium struggled to find a secure foothold in the marketplace. No single new 'knock-out' application had appeared, such as armour plating had in the case of nickel. To make headway it had to displace other metals, and so the producers, especially the Pittsburgh Reduction Company under the driving leadership of Arthur Davis, became skilled marketers.[14] University students were despatched door to door, selling to housewives the benefits of aluminium pots and pans. A thriving trade was developed in lightweight teapots. The attention of US Army quartermasters was directed to the possibilities of lightening the burdens to be carried by their infantrymen, and eventually military equipment such as canteens and tent pegs became a major market. Another market developed in the youthful automobile industry when it was realized that the ease of casting the new metal and its light weight allowed it to be effectively used in many different engine parts. Other uses were found, ranging from fishing-boats to reflectors for locomotive headlights. As customer orders grew, prices fell, and by the turn of the century aluminium was cheap enough to be able to challenge copper as the raw material for high-voltage electrical cables. One contributor to the fall in costs of production was the Bayer process for the production of alumina, the basic feedstock for the Hall–Héroult process. It had been perfected by an Austrian chemist, Karl Joseph Bayer, when working at a dye factory in St Petersburg. So well adapted was the process to treating bauxite that it was installed in British, French and German plants before the turn of the century. Alcoa adopted it sometime later.

World War I took aluminium to new heights.[15] When hostilities broke out, statesmen and generals had no experience of conflict between industrially advanced combatants. The generals and their staffs had been trained to appreciate the military value of the railway, but failed completely to comprehend the opportunities provided by the two new transport technologies of the automobile and the aeroplane. It did not take long for the learning to begin. A great massing of armies took place at the beginning of the War, and the rail systems of Europe were pressed into service to transport troops to the Front. But what if the Front was in an area not serviced by a railway? General Joffre, commander-in-chief of the French forces, was faced with just such a situation early in the month of September 1914. To halt the German advance that was now endangering Paris, he needed soldiers on the Marne River at a place where no railway came within 20 miles. Making matters worse, his troops were in Paris, and rail access to the region was cut. From this emergency was born the idea of using the taxis of the French capital to act as troop transports.[16] For three days an unending stream of sedan cars plied the muddy roads between Paris and the Marne. In the ensuing battle the German advance was stopped, and Paris saved. It was the first effective military use of the automobile. Likewise, the military potential of aircraft was soon appraised. At the beginning the biplane was used, along with hot air balloons, to spy out the disposition of enemy troops. Gradually it dawned on the army general staffs that not only could it be used to collect military intelligence, it could also be used to shoot down enemy aircraft engaged in similar missions.

With such a contribution to make, the automobile and aircraft factories were soon considered strategic industries of the greatest importance to the war effort. Being a vital material in the automobile engine, and an even more vital material in the aircraft engine, the demand for aluminium leapt upwards. The statistics tell the story. Global production rose by two-and-a-half times during the course of the conflict. In North America it increased more than threefold, while in Germany, where pre-war production was negligible, during the last year of the struggle a quantity second only to that of the United States was produced. Indeed, aluminium was of particular importance in Germany, where the scarcity of copper provided that country's metallurgists with the impetus for the development of stronger aluminium-based alloys, and also for the perfection of a welding process that allowed use of those alloys in a far wider range of applications. Hugo Junkers built aircraft for which the fuselage and wings were constructed out of a tough corrugated aluminium-copper alloy known as Duralumin.

The understanding of aviation technology had been at the forefront of wartime technical research. As a result, average air speeds had increased and in-flight stability and manoeuvrability improved. The basic design had evolved more slowly, and at war's end most aircraft were still biplanes. Typically they were constructed of a welded tubular steel frame or, more commonly, a wooden frame held together with glue. The fuselage and wings were covered with a mix of plywood and fabric. The limitations of wood and fabric were well known, and dissatisfaction was growing. Wood was prone to cracking in mid-flight, particularly as speeds increased and stresses became more severe. It was a weakness that became more pronounced after the wood became wet. Fabric also presented problems at high speeds, not always being able to resist the air drag across the wings.

Dissatisfaction was also developing with the unwanted air resistance created by the double-decker wings and struts of the biplanes. For these reasons designers began to intensify their experiments with all-metal single-wing aircraft.[17] The eventual victory of this design took some time. The monoplane was more difficult to control, and the metal coverings made it heavier and much more expensive. One of its main attractions was its ability to sustain the stresses of higher speeds, an ability that was of little use, however, if higher speeds could not be attained. The early metal aircraft found it difficult to outperform the wood and canvas models in this respect.

Efforts continued, often encouraged by the army and navy top brass, many of whom were convinced that the sleek metallic bird-like designs were the future of flight. Gradually experience accrued until at last, in 1930, the all-metal O-19 built by the firm of Thomas-Morse sufficiently impressed the US Army to place an order for 130 of them. The aluminium-skinned monoplane had become the standard in one of the largest air forces in the world. Not long after this, the fate of the wooden aircraft was sealed when the US Army decided against the planting of a spruce reserve that would guarantee adequate supplies of wood suitable for aircraft construction. Adolf Hitler provided the next boost for the metal airplane, when he assigned to Herman Goering the task of developing a modern air force. The Reich's Air Marshal turned to the Duralumin-skinned prototypes of the German designers Junkers, Heinkel and Messerschmitt. The Luftwaffe set the benchmark for military aircraft. It is estimated that, in 1933, throughout the world, a total of 4,000 warplanes were assembled. In 1940 this had increased to nearly 40,000, and the assembly lines were just warming-up.[18] In the next five years the demands of global warfare would create an insatiable appetite for fighters, bombers and airborne troop transports of all shapes and sizes. The consumption of aluminium leaped, and then it soared. Before turning our attention to the second great European conflict, let us briefly examine the structure of the industry that grew to supply the metal that would underpin the war of the air.

The worldwide production of aluminium was from the start concentrated into a handful of companies. This was because the rights to the Hall–Héroult process were not widely distributed. In the United States, Charles Hall joined forces with a syndicate to form what would become the Aluminum Company of America. Paul Héroult was a little more liberal.[19] After forming his own firm, Société Froges, he licensed his process to the Swiss-based Aluminium Industrie Aktien Gesellschaft, or AIAG, which in turn built smelters that took advantage of the hydroelectric power available from the rushing rivers of the Alps. Then, in 1895, Héroult also sold the licence to some English entrepreneurs, who established British Aluminium. Fifteen to twenty years of patent protection gave these first movers the time to master the technology and lock up the known supplies of bauxite.

Despite the complexities of the process, they learned fast. Costs fell, and larger plants were built. As early as 1901 they found it necessary to gather together to form an international cartel in order to control the plunging price. The cartel lasted seven years before collapsing under the weight of a global recession and competition from new entrants in Italy. In France, the fall of the patent wall had allowed two new companies, Compagnie Alais and the Société d'Electrochimie, to join Paul Héroult's original firm. As a result that country for several years experienced something approaching genuine competition among rival producers. This was brought to an end when, in 1911, they banded together to form a domestic cartel called L'Aluminium Français. Later, two of them would merge to form Pechiney. (Its formal name was Compagnie Alais, Froges et Camargue, but it had long been known by the name of the man, Alain Pechiney, who had refused to take the firm into aluminium production.) Another new European firm appeared when, in order to manage their crash programme of smelter construction, in 1915 the copper-deprived German government created Vereinigte Aluminiumwerke (United Aluminium).

After World War I, the Aluminum Company of America (Alcoa) and the five biggest European firms banded together to form the Aluminium Alliance.[20] It was a model of orderly global cooperation, one so successful that on the eve of World War II, if we exclude the Soviet Union, we find the same handful of companies still firmly in control of global production. Easily the largest of them was Alcoa. After 50 years it still stood alone on the North American continent. Political considerations had forced it to put some distance between its United States operations and those in Canada, and to that end it had created a nominally autonomous but still fully-owned Canadian subsidiary named Aluminium Limited. In addition to the Niagara Falls smelter, Alcoa and its northern sibling controlled other smelting works in Canada and still others south of the border. The group owned a large alumina refinery in Kansas, on the banks of the Mississippi, and fed it partly from the produce of mines in Arkansas and Georgia. The depletion of these reserves had prompted the development of the bauxite of Dutch Guyana that, along with neighbouring deposits in British Guyana, had come to supply the best part of the company's requirements. In an eloquent testimony to the strength of the American instinct for protecting its own industries against the competition of outsiders, the US Congress felt it necessary to protect this monopolistic industrial behemoth by maintaining a tariff wall to keep out the Europeans.[21]

The largest of the European producers was Vereinigte Aluminiumwerke.[22] In the hands of the Nazis it had become an important part of their rearmament and reindustrialization programs. Second in size was AIAG. Owned and financed by German industrialists, it also directed the bulk of its output to German factories. Both firms relied on bauxite railed in from deposits in Hungary that had been opened up in 1927, when, prompted by restrictions on access to Italian supplies, Vereinigte Aluminiumwerke had embarked on a search for new sources. Close behind the Germans was Pechiney, based in southern France and still mining the ores of Les Baux. Among the members of the cartel close relationships had developed, one example being the three-way shared ownership between Aluminium Limited, British Aluminium and Pechiney of Norway's Norsk Hydroelectric smelting works. But political events were once more shoving aside business considerations. In 1936, stretched beyond its limits by the tensions accompanying German rearmament, the Aluminium Alliance broke down.

The preparation for war added a new dynamic to global metals. Joffre had appraised the situation well. The spiteful peace had left Germany weakened, humiliated and unstable, and the French army had occupied the Saar coalfields in retaliation for German slowness in paying the war compensation. After more than a decade of bickering, the divisions within the German body politic were such that Adolf Hitler, funded and encouraged by a handful of wealthy industrialists led by the steelmaker Albert Thyssen and including the Krupp family, was able to gain over 30 per cent of the popular vote at the 1932 Reichstag elections. The next year he was Chancellor, and initiated a far-reaching programme of rearmament and public works. Though rearmament was expressly forbidden by the terms of the peace, no nation could muster the strength of will to call Hitler to account. In Britain, Winston Churchill thundered in Parliament against the growing menace, but his political stock was too low and no one wanted to hear such disturbing talk.

In 1937 Hitler's revitalized armies reclaimed the Saar, sweeping aside the French as if they were cardboard cutouts. In the same year nearly two-fifths of German steel[23] and over one-third of its aluminium were devoted to military uses. The country's total aluminium production had surpassed that of the United States. Swedish iron ore now became so vital a staple of the rearmament programme that Hitler is said to have mused that his first step after declaring war would be to occupy that country. The rest of the world had it within their power to prevent, or certainly to retard, the German build up. Yet ships laden with Canadian nickel, tin from Malaysia and Rhodesian chrome continued to dock at Germany's Baltic ports. No one doubted that their cargoes were grist for the military machine. Throughout the decade Hitler's foreign policy pushed closer to the brink, scorning the French and British who tried to accommodate his demands. In Soviet Russia, Joseph Stalin signed a non-aggression pact with his mortal enemies. When German tanks rolled into Poland in 1939, even the most determined advocate of appeasement had to admit that Joffre's ceasefire was broken beyond repair. War was declared once more between the industrial nations of Europe.

This time there could have been few among the well-informed who would have expected a lack of metals to bring Hitler's armies to their knees. Nevertheless, there were those who hoped. One was the French Prime Minister, Paul Reynaud. 'The iron route is cut and will stay cut', he declared in May 1940 after the Allies had taken the Norwegian port of Narvik, the shipping base for Swedish iron ore.[24] It would take more than wishful thinking to cripple German might. The iron ore kept flowing from ports further east, and a few months later even Narvik would re-open to German traffic as French troops were withdrawn in an attempt to repel the tanks that were crashing through the forests of northern France. Nickel, too, came from the north with the expansion of the Finnish deposit of Petsamo and the capture of a Norwegian refinery. The fall of Hungary netted the Nazis a good supply of bauxite, and the invasion of the Ukraine enabled Hitler to occupy the largest manganese fields in the world. Thanks to stockpiles accumulated during the build up to war, there was even an adequate supply of those scarcer metals vital to the manufacture of armour-piercing shells and specialty steels.

When the Japanese entered the fray, their lightning conquests into south-east Asia secured them control of much of the world's tin as well as vital supplies of rubber, petroleum and iron ore. In a gloomy speech given to the American Zinc Institute in 1942, the head of the Economics and Statistics Branch at the US Bureau of Mines detailed the growing mineral wealth of Germany and Japan. By his reckoning, where once the aggressors had held a mere twentieth of global mineral resources, they now controlled more than one-third.[25] Grim as the picture appeared, the fact was that the eventual outcome had already been decided. The bombing of Pearl Harbor was the turning-point. On the surface it was a brilliant Japanese victory, but the real significance of Pearl Harbor was that it forced an isolationist United States into the war. It would take another three years before the Allies would be certain of victory. Even then, the Japanese continued the struggle until, on 6 August 1945, a bomb of awesome power was dropped on the city of Hiroshima, and three days later another devastated Nagasaki. Contained within each was eight kilograms of a metal that only five years before had been looked on as waste rock.

The first step taken on the road to the atomic bomb was made on a day in 1895 when Wilhelm Roentgen, a professor of physics at the University of Wurzburg, was passing some electric current through a vacuum tube. When he did so he noticed that a piece of metal some yards away began to glow. It continued to glow when he placed paper, then wood and finally aluminium between the tube and the metal. Puzzled by this phenomenon, he named it X-radiation. The X-rays left images on photographic plates. They also passed easily through human flesh, less easily through bone. In time the professor would amuse himself by taking photographs of the bones in his wife's hand. The enigmatic X-rays captured the scientific imagination. Rays of all kinds – even those from glow-worms – became the vogue field of study. One physicist so intrigued was Henri Becquerel. He conducted a series of experiments during which he demonstrated that the heavy metal uranium emitted something very similar to X-rays without needing any electrical stimulation. Uranium was, to use the word coined by Marie Curie, radioactive. Becquerel penned a few papers, and quietly forgot his curiousity. The baton passed to Marie and Pierre Curie, a young couple working in the laboratories of the Ecole Polytechnique in Paris.

The Curies set to work on a material known as pitchblende. It emitted stronger rays than pure uranium. There must be, Marie hypothesized, some other substance within it. The quest to isolate radium, as the substance would be called, would win for her a Nobel Prize for Chemistry. Radium had the unique property of being wonderfully strong as a source of radioactivity. And radioactivity was more than ever the fascination of the scientific world. It promised to be the key that would unlock the secrets of the atom. Essential experimental work could not be done without radium. Demand for it increased when other properties became known. Its rays could destroy cancer cells and, when painted on watch dials, it glowed in the dark. It became a coveted material. It could not, therefore, have been a difficult decision for the directors of

Union Minière when presented with the discovery of a rich deposit of uranium near the village of Shinkolobwe in Katanga.[26] No one considered the uranium itself to have any value, but as every geologist knew, where uranium lay, there too lay radium, admittedly in very small amounts. It required three tonnes of uranium ore to produce a single gram of radium. Although this made the cost of mining and processing unusually expensive, the price made it all worthwhile. Weight for weight, radium was about 20,000 times more valuable than gold. The decision to develop Shinkolobwe was duly made, and in 1921 a mill was built in the Belgian town of Oolen near Antwerp. There the uranium of Katanga was stripped of its precious content of radium and then dumped just outside the town. The yellowish waste was left to soak up the rain.

In the meantime, science progressed. Ernest Rutherford and his brilliant band of young physicists showed that the atom was mostly empty space in which a handful of electrons flashed in orbit around a tiny core. Albert Einstein published his relativity theory, part of which consisted of a demonstration that mass could be converted to energy. One of Rutherford's students, Niels Bohr, invented the concept of quantum mechanics. Werner Heisenberg, working with Bohr, developed the 'uncertainty principle', thereby forever dispelling Newtonian notions of causality. The first particle accelerator allowed two more of Rutherford's men to undertake experiments that involved hurling atomic particles at a block of the metal lithium. The collision of the particles with the lithium actually broke the lithium atoms into two. The atom had been 'split'. The more sophisticated term for this was nuclear fission. Then, in 1930, James Chadwick, yet another protégé of Rutherford, found the elusive neutron. Step by step the secrets of the atom were laid bare.

Amid all the excitement Rutherford foretold the development of terrifying bombs that used the energy of the atomic nucleus. Two German physicists, Hahn and Strassman, succeeded in unleashing this energy when they managed to split atoms of uranium using neutrons fired through a particle accelerator. Enrico Fermi suggested that neutrons released during the uranium fission would cause still other atoms to split, thus creating a chain reaction. Scientific speculation turned anew to the possibility of harnessing this source of energy. War broke out that same year. In the spring of 1940 German weapons specialists arrived at Oolen for a talk with the management of the radium factory. The Germans were not alone in identifying the military potential of nuclear fission. In 1939 Albert Einstein was persuaded to write a letter to President Roosevelt urging him to fund studies into the creation of a fission bomb.

The lead up to war had seen many of the world's most accomplished nuclear physicists emigrate to the United States, and the centre of this young field of research came to be Columbia University in New York's Manhattan. Soon enough their attentions came to focus on the possibility of a bomb. The Manhattan Project, under the leadership of Brigadier-General Leslie Groves, would eventually become a massive military-run programme with research facilities throughout the United States. A thought must be spared for General Groves. He was a military man accustomed to managing civil construction projects, and he found in the scientific community a group entirely alien to his experience. A clash of cultures ensued. Probably because he was the only scientist with whom Groves could talk sense, Robert Oppenheimer was appointed to run the technical end of the Project. Leaked information suggested that their enemies were hot on the trail of a workable bomb, and fear of losing the race spurred on the American effort.

The most difficult task facing the small army of scientists was to separate the relatively inactive uranium-238 from the highly fission-prone uranium-235, thus providing fuel for the bomb. (Uranium occurs naturally in these two isotopes – uranium 238 and uranium 235. The ratio is about 100:1 of 238 to 235. The two were separated by heating them to gaseous state. U-235 passes slightly more easily than does U-238 through a very fine mesh screen.) The technical solution was found in 1943, and on the banks of the Columbia River in Washington State was built a uranium concentration plant. Supplying the plant with uranium was another challenge. The only known sources of the metal were Shinkolobwe and the smaller deposit of Great Bear Lake in Canada. In 1939 the British had offered to purchase the entire output of Shinkolobwe. While the request was refused, a year later the fear of a German invasion of the Congo prompted Edward Sengier, head of Union Minière, to pack into drums all the uranium ore sitting stockpiled at the mine and ship it to New York.[27] There it sat gathering dust in a warehouse for two years before American authorities finally woke up to its significance.

Robert Oppenheimer's first move was to establish a new facility in which all the disparate nuclear research projects could be brought under the one roof and driven hard in military fashion. The site chosen was the New Mexico town of Los Alamos. There the war came to life only in newspapers and in the reports of new arrivals, many of whom, like Neils Bohr, had been smuggled out of occupied territory. It was in Los Alamos in July 1945 that the world's first atomic bomb was assembled for testing. On the morning of 16 July it sat perched on top of a steel tower located in an expanse of New Mexico desert near the Alomogordo air base. The group of observers, secure in their concrete bunker several miles away, probably felt some tension. Would it work? What would happen? At 5:30 am they got their answer. As the clock struck the half hour, a searing wave of heat swept across the desert, leaving the steel tower vaporized and the desert sand fused to glass. Then followed the thunderous noise of the blast. A huge mushroom-shaped cloud billowed into the upper reaches of the atmosphere.

By this time uranium had become a metal of the highest strategic importance. Groves was ordered by Roosevelt to achieve as complete control as possible of the world's reserves. His secret survey showed Katanga to be the main, and easily the largest, source. The second most important was the mound of waste at Oolen. Groves could only assume that this latter supply had fallen into Nazi hands. He was right – the stockpile at Oolen had been trucked to Germany. What Groves and his team did not know was that the German atomic bomb programme never really got off the ground. Indeed, any hopes for a Nazi programme on the scale of the Manhattan Project had ended in late 1942 when Werner Heisenberg, in a presentation to Albert Speer, could muster only the most lukewarm assessment of the prospects for the successful development of atomic weapons. Controversy has surrounded Heisenberg's role ever since.[28] Was he merely incompetent, or did he pour cold water on the project for fear of the weapon falling into Nazi hands? Whichever was the case, Heisenberg and his colleagues had singularly failed in their task. Near the end of the war Groves sent in a team with orders to seize the Oolen uranium and the German researchers. They traced Heisenberg to the south of Germany to the small town of Hechingen. There they found in a cave just out of town a primitive nuclear reactor, the height of German success in its pursuit of the atomic bomb. The yellow ore was found nearby in a flimsy barn. It was trucked across the border out of reach of the West's next enemy – Stalin and his Red Army.

A CHALLENGER EMERGES

To outsiders, the Soviet Union was something of an unknown quantity, its huge reserves of metallic ores in the Ural Mountains and Siberia being off-limits to the Western mining men of the 1940s. It had not always been that way. Following the great flowering of metallurgy in the American West, the eyes of the world's miners had turned to the four corners of the earth in search of fresh fields. The attention of some had come to rest on the 'sleeping giant' of Russia. Its potential had long been known. Had notTocqueville declared that the United States and the Russian Empire would be the dominant world powers of the next century? By 1900 the United States was well on the way to fulfilling his prophecy; in Russia the conservatism of the Orthodox Church and the landowning class lay like a heavy blanket over the economic landscape. Not all of the ruling class were content to let their country moulder while the West grew each year in power and prosperity. Count Witte was for a time the leader of this group, and in his hands economic reform in Russia received a strong boost. He championed the construction of the Trans-Siberia railway, converted the currency to the gold standard, and opened the doors to European industrialists willing to invest money in Russian factories.

One of the focal points of interest for foreign capital was the coal- and iron-rich Donbass, a region that lay just north of the

Black Sea in the triangle formed by the Don, Donets and Dnieper rivers.[29] Belgian, French and British steelmakers had established mines and blast furnaces there, creating the foundation of what later would become one of the main supporting pillars of Soviet industry. One of them was a Scottish engineer named John Hughes. In 1874 he built a steel rail factory, and in so doing gave his name to the settlement that grew up around it. Over time Yuzovka would become one of the metallurgical centres of the region. Other European consortia exploited the oil deposits around Baku on the Caspian Sea. Among them were the Nobel brothers and an Englishman by the name of Leslie Urquhart. Later, after losing a fortune in Baku, Urquhart would earn himself the sobriquet of 'the Mining Czar' for the range of his interests, which extended from the Urals to Kazakhstan.[30] Americans were at the back of the pack. At one point the ubiquitous John Hays Hammond led a delegation of businessmen intent on securing mineral rights to whatever valuable properties they could identify. Aside from affording Hammond the chance to lecture the Tsar on the basics of economic and social policy, the trip was a failure. That a trip like this failed was not unusual. The advance of Russian industrialization had suffered a blow when Witte was assassinated in 1905. The ruling class fell back into its old habits of repression and self-indulgence. By the outbreak of World War I, mining and smelting activity in the whole of Russia amounted to not very much.

Discontent with this state of affairs boiled over when it became apparent how incompetent the military leadership of the Tsar and his generals was. Eventually the Bolsheviks, led by Vladimir Lenin, took advantage of the confusion and the leadership vacuum and seized power in 1917. Lenin's two strongest policy convictions were the need for an organization that would help foment Communist revolutions across the world and the need to establish a strong industrial base within his own country. He had little use for the levers of conventional economic management. Gold, he said, would be best put to use in filling teeth. Whatever remained might as well be used as plumbing fixtures in public lavatories. Copper, on the other hand, would be needed in huge quantities to help construct the network of high-voltage power lines that would bring Russia into the industrial age. In the first flush of revolution the Bolsheviks ceased payments on all the Tsarist international borrowings, and, once the foreign capitalists had departed, nationalized all foreign-owned enterprises. They then set about creating a Russia of factories and equality. Revolutions do not succeed that easily, and progress stalled almost immediately. Counter-revolution and then compromise led to a greatly slowed pace of change, and at the end of the first decade of Russian communism the electrification of the country remained a pipe-dream and conditions for the majority were probably worse, and certainly no better, then they were under the Tsars.

After the death of Lenin in 1924, Joseph Stalin (the name means Man of Steel) won the struggle for power within the Bolshevik ranks. Declaring himself in favour of 'Socialism in one country', he abandoned the efforts to create global revolution and instead focused his energies on strengthening Russia. To him there was no possibility for doubt: industrialization was the key. He determined that it would be achieved through a series of Five Year Plans. The first, which came into effect in late 1928, called for huge increases in all manner of heavy manufactures. Steel production was targeted to increase in the five years from 3 million tonnes to 16 million. Copper output was to grow fivefold,[31] coal merely to double. There was even renewed interest in gold, for the Soviet government had discovered that it needed the outside world. It needed foreign engineers and foreign machinery. While both were readily available, they had to be paid for in hard currency. Gold was known to exist in the uninhabited forests and steppes of eastern Siberia, and on this Stalin set his sights.[32] Exploitation of these deposits would have its strategic advantages too, as the gold-bearing regions included the territory abutting Manchuria, the northern Chinese province that was at that time under the occupation of the expansionist Japanese. Searching for a way to secure both the gold and shore up his eastern flank, Stalin, so the story goes, found the answer after reading several accounts of how the Californian Gold Rush had led to the populating of the western United States. He wanted the same thing in the east.

The ambitious, to say the least, Plan targets were not based on what technical men considered possible; rather, they were based on what Stalin considered necessary. He expressed his thoughts in a 1931 speech to a conference of factory managers:

> To slacken the tempo would mean falling behind. And those who fall behind get beaten. But we do not want to be beaten. No, we refuse to be beaten! That is why we must no longer lag behind.... We are fifty to one hundred years behind the advanced countries. We must make good this distance in ten years. Either we do it or we shall be crushed.[33]

The task set was a huge one. For example, at the time of Stalin's speech, construction was just nearing completion on the country's first aluminium smelter. As was typical of the Bolshevik style, the Volkhov smelter was a built on a grand scale. Located near Leningrad, it used the ores of nearby Tikhvin. From the start the project was beset with problems. On the one hand, the ores were of such poor quality that technical men from Alcoa, twice invited to inspect the deposit, had both times pronounced it unsuitable for use in alumina refining. Furthermore, the Soviet engineers had no experience with the difficult technology aside from what they had gained during laboratory testwork. On the other hand, planners in Moscow, deaf to any pleas for patience, applied relentless pressure for more and more production. If achieved, their demands would have meant that by 1938 the Soviet annual output of aluminium would have surpassed the United States. The result was chaos accompanied by a steady progress.

The same pressure was applied to every other industrial metal. When it was realized that known ore deposits were insufficient to supply the Plan targets, Vasily Gulin, head of an organization that managed Urals copper production, was appointed to lead a Geological Trust that would drive mineral exploration throughout the Urals, Kazakhstan and elsewhere.[34] The Trust achieved considerable success in finding a range of copper, bauxite, nickel and lead and zinc deposits, and thus provided the feedstock for the construction of a series of large state-owned industrial complexes that were the foundation of the Plan. An aluminium refining and smelting plant was constructed in the northern Urals town of Kamensk, and

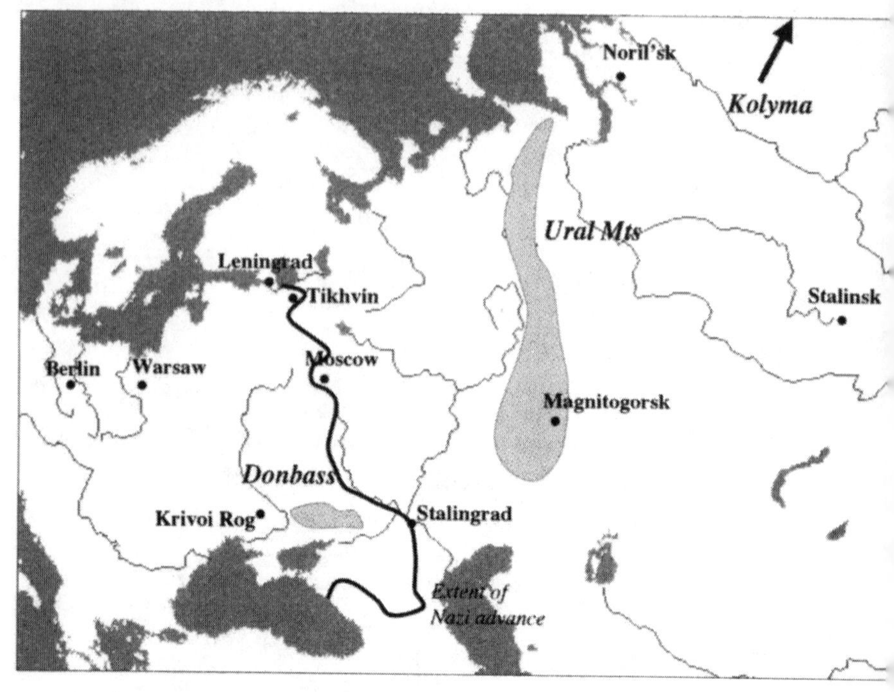

11 Soviet Russia, 1938, with the final reach of the Nazi penetration (1943) added

that mountain range also became home to so many other base metal operations that by 1938 it was one of the premier metal regions in the world. It was also home to the crowning glory of Stalin's programme. For here was located the steel city of Magnitogorsk.

The expansion of the steel industry was the centrepiece, almost the purpose, of the first Five Year Plan. Two new iron mining and smelting complexes were to be constructed from the ground up. The first, Magnitogorsk, would be built near Mount Magnitaya, a rich mountain of iron.[35] The second, at Stalinsk in central Siberia, was to be located near great reserves of coal. A railway more than 1,500 miles in length would join them, carrying ore one way and coal the other. The sites were chosen for more than just their good mineral reserves. It was an aim of Soviet development that the poorer eastern regions would share in the benefits of industrialization. More importantly Magnitogorsk and Stalinsk were both distant from

Russia's western frontier, which, long history had shown, was vulnerable to attack from her European neighbours. Not everyone appreciated the strategy. Stalin had not yet attained the absolute and despotic power that he gained in later years, and the massive scale of the project – it would create at Magnitogorsk the largest steel production complex in the world – came under heavy and persistent criticism. Most vocal were the Ukrainian steelmen, who considered it an absurd undertaking when a similar expansion in their own Donbass could be achieved far more quickly and at much lower cost. Stalin forced the project through. In later years he would have his revenge on the 'oppositionists'.

All the energy, enthusiasm, haste and waste of Stalin's crash industrialization programme was reflected in the building of Magnitogorsk. The first design for the steel city was drawn up by a Soviet team. It was deemed inadequate, and an American firm, Arthur McKee & Company, was commissioned to revise the plans and advise throughout the construction.[36] The Americans were appalled at the expectations of the Soviet high command. The timetable would have been ambitious even if the project was being undertaken with a skilled, well-equipped workforce located near industrial infrastructure. Instead it was in the middle of nowhere, and the workforce totally inexperienced. Their objections were overridden. Arguments ensued, designs were changed and timetables slipped. The site was still a wilderness and the overall layout still in dispute when, in early 1929, thousands of young volunteer workers flocked in to begin construction. The first McKee engineers reached the site a year later. There they found chaos. The Russian side of the management team was being changed at regular intervals, sometimes for reasons related to the construction and as often because they had fallen foul of the political commissars who were assigned to ensure unwavering loyalty to the Stalinist line. For their part the Russians considered their foreign counterparts to be cowards when, after voicing their criticisms, the Americans refused to take accountability for putting the project to rights.

Everything from food to lumber to labour was in short supply. The approach of a trainful of lumber would prompt a flurry of telegrams to Sergo Ordhzhonikidze, the Minister for

Heavy Industry, and even to Stalin himself, seeking their support in laying claim to the precious material. Amid a chronic shortage of manual workers were buildings full of office staff whose task it was to keep track of every detail of the project. Inexorably the huge edifice took shape. A final hectic push resulted in the first molten pig iron being poured in February 1932, although the flow lasted only days before the furnace was shut down for major repairs. From then on the output steadily grew, fast enough to amaze the resident foreigners, but never sufficiently to satisfy the targets set in the Five Year Plan. Later on, as the targets still remained out of reach, a new phenomenon appeared. This was the Stakhanovite movement, named after a Donbass coal miner who, in 1935, dug in a single shift some 100 tonnes of coal, when the normal performance was less than ten. It showed, trumpeted the Communist Party dailies, what a true Soviet man could do when the need arose. It also justified the authorities in raising the work output demanded of their already hard-pressed labour force.

The Five Year Plan targets were too ambitious even for the iron will of Stalin and his Stakhanovite 'shock workers'. The first Plan targets were more or less achieved only by 1940. That they were reached even then may be considered a remarkable feat, given the turmoil that raged throughout the country during the intervening years. For Stalin, at the same time as launching his Five Year Plan, had decided to tackle the problem of agriculture. It had always been a Bolshevik aim to industrialize the farming sector through the creation of 'collectives', large rural farms equipped with tractors and other machinery that a small farm of the old type could never afford. Not only would the collectives improve efficiency and release workers who could then be employed in the industrial cities, they would destroy the old rural order. One obstacle stood in the way: the whole idea of the collectives was anathema to the peasants, and especially to those who tilled the fertile plains of the Ukraine. They refused to join them, and in the ensuing struggle perhaps 10 million people died through famine and hardship.

The damage done to agriculture would leave Russia hungry for decades. Entire farming communities were deported to the factories and mines of the Urals, Kazakhstan and Siberia. It was decided at the same time that the nomads of Russia's Moslem republics had also to be dragooned into the Soviet system. Many were employed in the scattered and isolated gold mines of Kazakhstan and Siberia. An American mining engineer, John Littlepage, described the scene:

> It can be imagined what a heartbreaking task it was to teach such workmen to use air-drills, modern milling equipment and especially to handle dynamite. I don't understand yet how they failed to blow themselves up and all the rest of us. My own worries on this subject were not quieted when I went into a bath-house one day and found a group of them bathing with cakes of cyanide which they had mistaken for soap.[37]

Littlepage fails to mention that the fact that they were bathing at all was an unmistakable sign of progress in assimilation.

The inevitable failures and disasters that accompanied the industrialization programme led to allegations of sabotage which in turn led to round after round of arrests and deportations. No doubt many of the allegations were well-founded. After all, the extent of Stalin's tyranny could hardly fail to excite determined opposition among the embittered peasants, dispossessed nomads and expelled Communist Party members. He was hated in the Ukraine. In his account of his nearly ten years work in the gold and base metal mines of the Stalin era, Littlepage tells of an incident at the Ridder lead and zinc mine near the border with Mongolia. He had spent, he recalled, some time rectifying various technical problems at the mine, and then left behind detailed instructions for further work with two young engineers whom he considered very capable. On his return he found the instructions flagrantly disregarded, the underground works in a mess and the pair under arrest for sabotage. Elsewhere he tells of gearboxes full of sand and deliberately shoddy workmanship.

The annual tally of arrests swelled into the millions after 1935, when the Russian dictator began his purge of the Communist Party. Gulin disappeared, a victim of the over-ambitious copper production targets, and the popular Orzdhonikidze, a longtime ally of Stalin, committed suicide. Shevchenko, the manager of the coking section at Magnitogorsk, and his 'gang' all received heavy sentences for sabotage following mishaps at the plant. John Scott, an American mechanic working on the project, considers some of the accused to have been guilty and others guilty only by being in the wrong place. The engineers at Tikhvin were arrested, accused of treachery and sabotage by the Moscow-based Director of the Institute of Non-ferrous Metals. They were just drops in the ocean. In his memoirs, the former Donbass miner and Stalin's successor, Nikolai Krushchev, refers to the 'meat grinder of 1937'.

It was a dark time. A never-ending stream of Ukrainians, Crimeans and Chechens as well as expelled Party officials and other 'saboteurs' found themselves packed onto rail wagons and hauled to the east Siberian city of Vladivostock. From there they were crammed into prison ships and ferried north to the town of Magadan, the wind-battered entry port to the frozen wastes of the Kolyma Peninsula.[38] Those who had not died on the journey were marched inland wearing only whatever clothes they had managed to grab during their arrest. On arrival at their shelterless campsite they were set to work digging gold. For gold there was aplenty in the rivers and lakes of the peninsula. The first gold seekers to arrive in Kolyma, sometime during the mid-1920s, were free prospectors working on the fringes of the law, and for awhile thereafter the government had toyed with the idea of harnessing their energy by promoting a California-style rush. But then the Central Committee of the Communist Party had a better idea. They would force into service the stubborn peasants of the Ukraine. Thus began the policy of sending prisoners to work as gold-mining slaves in the sub-Arctic wilderness until their inevitable death from starvation, cold or disease.

It was a merciless existence. The prisoners were set a daily quota of production, and failure to achieve it meant reduced rations or a trip to the punishment cell. Even the standard rations were barely enough to survive on, and few survived through three years of gold mining in Kolyma. One thorough estimate maintains that the goldfields claimed two million lives before the prison camps were closed in 1957. Amid the suffering a treasure chest of bullion was won, first by manual labour alone and later with the help of machinery. The peninsula became the cornerstone of a Soviet gold industry that in some years even rivalled in output the mines of the Rand. So it was that Stalin got his gold rush. Being frequently located in remote regions, many other mines were developed and worked by prisoners. The nickel and cobalt deposit of Noril'sk was carved out of the tundra by the doomed deportees.

The purges destroyed the momentum of the great achievements that were set in train by the first Five Year Plan. Building work virtually ceased at Magnitogorsk and elsewhere, yet there was no denying the magnitude of the progress. In the year before the Nazi invasion, Russian metal production had reached the level at which the United States had stood on the eve of World War I. In a catastrophe that vindicated Stalin's policy, a large part of this was captured soon after Hitler's armies marched into the Ukraine. Just ahead of the invaders worked squads of labourers whose task it was to dismantle as many factories as possible and rail them to safety. Equipment from the aluminium smelter at Volkhov was shifted 2,000 miles eastward and reassembled in Siberia.[39] Although the Americans shipped to Russia hundreds of thousands of tonnes of copper, aluminium and other metals, without the new steel cities and the rest of the Stalinist metallurgical developments, the Soviet Union would have struggled to manufacture sufficient tanks, planes and ammunition to fight and win the epic battles for Stalingrad and Moscow. These victories halted the German advance, and by 1944 the invaders were in full retreat. Russian battalions were the first of the Allied forces to enter Berlin.

THE POST-WAR BOOM

The Second World War was by far the most destructive in history. Europe and Japan lay in ruins, their cities and factories demolished by the aerial saturation bombing campaigns of the final stages of the struggle. In Nazi-occupied Russia the tank and artillery battles had wrought similar damage. In North America, the United Kingdom and the Urals alone in the world could be found major industrial centres still intact. The damage was such that, in 1946, the United States was estimated to account for more than one half of the world's entire industrial output. The question on everyone's lips was: how best to rebuild? Should the defeated aggressors be permitted to rebuild the heavy industry that might allow them once more to wage war? How should the United States become involved in the tasks of reconstruction? Should the Soviet Union take advantage of American aid? As politicians, generals and bureaucrats searched for a path forward, two developments had by 1950 put an end to most of the debate. First, Stalin refused American help and turned his weapons against his former allies. The Soviet intent was sufficiently clear by 1948 for Winston Churchill to be able to say that 'From Stettin in the Baltic Sea to Trieste in the Adriatic, an iron curtain has descended across the continent'. Then, in 1950, the Korean War kick-started a global economic boom that would last the best part of 25 years.

The boom had its origin in the renewed military demands of the war in Asia, and it was borne along, particularly in North America and Western Europe, by a sustained surge in popular demand for all sorts of light manufactured goods. This demand had been thwarted for twenty years, first by depression, then by war and then by the post-war reconstruction. Now the desire for consumer durables – refrigerators, the family motor car, washing machines, air conditioners and so on – grabbed hold of the economies of the advanced nations. An important contributor to the boom was the increasing affordability of such goods, partly the result of wartime advances in manufacturing productivity. A second factor sustaining the boom was the continuing build-up, even after ceasefire was declared in Korea, of military hardware and defence infrastructure, as the United States and the Soviet Union prepared themselves for the next great war. In the United States, successive Presidents authorized 'strategic stockpiles' of metals to be constructed in readiness for the time when the country would once more be plunged into global combat.[40] Stalin and his successors pursued a similar policy. As a result of all this, the world needed metals as never before.

The first response was an expansion of existing mining operations and a rapid adoption of the new technologies that could assist this. Rubber-tyred dump-trucks replaced the cumbersome rail spurs in open pit operations, and their flexibility and speed allowed larger tonnages to be moved. Underground, the new science of rock mechanics permitted the widening of tunnels and working areas. The outcome was a more spacious and stable underground environment that allowed, in turn, the adoption of mechanized mining techniques. The capacity of crushers and grinding mills grew larger, assisted by wartime advances in electric motor and mechanical power transmission technology. Improvements in the field of automatic control permitted more effective manipulation of metallurgical processes. A new technique in iron-ore processing extended the life of existing fields.[41] Pellets, as they were known, revived the Iron Range of Minnesota and did the same in the Soviet Union, where a similar depletion was affecting Mt Magnitaya and the long-worked iron-fields of Krivoy Rog in the Donbass.

A new twist on an old technique changed the economics of steel making.[42] Since the inventions of the Siemens brothers in the 1880s, the smelting of steel had remained unchanged in its essential steps. The Siemens open-hearth furnace had done its job well, with the result that the quality of steel had become predictable melt after melt, a substantial improvement on the Bessemer design. The drawback was that the open hearth was a lot slower than Bessemer's original process, because it relied on blowing air over the top of the pig iron rather than bubbling it through the molten bath. When Austrian engineers mastered the use of oxygen in place of air, a new technology, called basic oxygen conversion, emerged.[43] Oxygen did for steel manufacture what Cort's process had done for the production of iron – it greatly speeded it up. Not only could pig iron now be converted to steel in one-tenth of the time, compared to the open hearth a basic oxygen furnace cost half as much to construct and much less to operate. It opened the way for a new generation of steel mills of much greater size. In other metals, technological change was slower. In copper a coal-free smelting technique – flash smelting – was developed, but its adoption would be held up until environmental protection legislation changed the industry economics. In all sectors smelter size grew, and along the coastlines of Europe and Japan a new generation of giant metallurgical works were built to take advantage of the economies of scale now on offer.

Advances like pelletizing and an improved understanding of the mysteries of flotation may have extended the life of existing mineral fields, but the realization was growing that the future supply of ores for the industrial nations lay outside their borders and, mostly, across the seas. The point was forcefully made in the 1952 final report of the Paley Commission (the formal name being the President's Materials Policy Commission).[44] It rejected on two grounds a return by the United States to any insular policy of mineral self-sufficiency. First, the report stated, it was manifestly obvious that known domestic resources would be unable to sustain the American industry for much more than another generation or two – just look at the once inexhaustible Mesabi. Second, even attempting such a thing was inadvisable, now that the world was divided into two mutually hostile camps plus a bevy of uncommitted onlookers whose allegiance was up for grabs. Nations whose inequitable economic and judicial structures rendered their working classes sympathetic to Communism were often those that were mineral-rich. If the Americans were to contain Communism, they could not afford the luxury of tariff walls to protect their miners.

In Western Europe, too, policy makers and planners saw a future in which the bulk of their mineral raw materials had to be shipped in over the ocean. Continental reserves of base metals – copper, lead and zinc – were rapidly dwindling. This in itself was no new thing: Europeans had long been reliant on imports of these ores to bolster their domestic sources. Now, however, even staple minerals such as iron ore and coal were becoming seriously depleted, and the remaining deposits were increasingly either too low in grade or too expensive to work. With Hungary now locked away behind the Iron Curtain, even Continental bauxite was in short supply.

Then there was Japan. Once the outbreak of the Korean War had dissuaded General MacArthur from his plans to turn that country into a pastoral society, its economic revival had been swift. By 1960 the shores of Tokyo Bay were studded with the smoke-stacks of smelting furnaces, and important metallurgical centres had developed in Osaka, Kobe and many other cities. In comparative terms, Japan's need for seaborne mineral imports was greatest of all. This need can be exaggerated. The Japanese searched their island chain for mineral resources with such success that, even nearing the end of the boom, it still mined domestically nearly half of its requirements of zinc and lead, over a quarter of its copper and a respectable share of its coal.[45] Despite this, virtually all of its iron ore, bauxite and uranium, and of course a great quantity of every other mineral, had still to be imported.

Thus the United States, Western Europe and Japan all faced a situation in which a good portion of the ores and concentrates needed to feed their smelters would have to be shipped in. The question that remained unanswered was – from where? A worldwide search for minerals got underway. Equipped with their trusty binoculars, airborne geologists scoured the rugged ranges of South America and Africa. They also scanned the less accessible areas of Australia, Canada and Siberia. For such a task binoculars were clearly of limited value, and soon more sophisticated techniques were in use.[46] The first of them was the magnetometer, which proved itself particularly effective in locating iron ore deposits such as those found in the Canadian provinces of Labrador and Quebec. Aerial photography and infra-red scanning, both products of wartime aerial bombing campaigns, became other weapons in the arsenal of detection techniques. One technique, the Geiger counter, even became a household name when, during the 1950s, uranium enjoyed a brief period as the global glamour metal.

The activity revived Australia's moribund mining scene, which, after a few brief decades of glory during the heyday of the Copper Barons, had once again slipped into slumber. The torpor had been such that profitable production could not even be sustained at gold-rich Kalgoorlie. Now there emerged new leaders who were able to take advantage of the opportunities. Maurice Mawby took the reins of Consolidated Zinc, the descendant of Herbert Hoover's old company, the Zinc Corporation. Mawby's vision fixed on a broader horizon than that defined by the rooftops of Broken Hill. In 1955 he scored his first big success, when a team of company geologists spied during a prospecting flight across the northern tip of Australia a stretch of reddish coastal cliffs that on closer inspection turned out to be a massive bauxite deposit. When the evaluations were finally complete, it was realized that it comprised fully one-quarter of the world's known bauxite. Next he established a substantial cross-shareholding between his company and the prospect-rich but cash-poor Rio Tinto, which was busily investing in promising deposits across the globe.[47] It already held a stake in the Mary Kathleen uranium operation in the Australian state of Queensland. The first joint effort between the two companies took place in the early 1960s, when, in the dry and remote region of the Pilbara in Western Australia, huge high-grade deposits of iron ore were found. The geologists of Rio Tinto Zinc and Conzinc Riotinto Australia were at the head of the pack, striking an alliance with a local pastoralist named Lang Hancock and securing great swathes of territory from which would grow two of the world's largest iron mines. Some years later the companies developed a large copper and gold operation on the island of Bougainville, a part of the Australian colony of Papua New Guinea. Others also contributed to the Australian bonanza. Coal was found in Queensland, nickel and bauxite in Western Australia, manganese on an island not too far from Darwin, and in the Indonesian archipelago a breathtakingly spectacular copper mine and milling operation was carved out of the rugged peak of the Ertsberg mountain.

Finding minerals was not the only challenge. Bringing them to market could also prove difficult. Brazil presented a case in point. Since before World War I the world had known about the iron ore of the Minas Gerais province, and everyone knew of Percy Farquhar's three-decades-long series of fruitless attempts to persuade the Brazilians to establish on the back of these riches a domestic steel manufacturing and iron-ore exporting combine.[48] His efforts via the Itabira Iron Ore Company, which he had helped create in 1911, had every time foundered on the rocks of domestic political·rivalry and cronyism. During the war the British government bought out the Farquhar syndicate and handed the shares over to the Brazilians, who then proceeded to assume control of other iron ore assets and created the government-owned Companhia Vale do Rio Doce. In the immediate post-war period, still the iron remained largely undisturbed. Every attempt at upgrading and expanding the mine, rail and port facilities seemed to be tied to a sweetheart deal. At last in the 1950s a series of development-minded Presidents swept aside the political obstacles, and the rise to international prominence of CVRD was underway.[49] At the same time, the country was opened to foreign investment. Similar changes in the political climate took place in Peru, Venezuela and elsewhere. The resulting negotiations over access to minerals, the financing of their development and the conditions governing long-term sales agreements often pulled in high-level government officials on both sides. It was not at all uncommon for American firms operating in Latin America to appoint former senior members of the US government as their chief local officials. The practitioner par excellence of this mode of influence was the firm of M. A. Hanna. Employed in its Brazilian office to help the company secure a foothold in Minas Gerais was a truly star-studded line-up. There was George Humphrey, former Secretary of the Treasury in the Eisenhower Administration, and John McCloy, a former President of the World Bank. Walk-on roles were assigned to the sons of Herbert Hoover and John Foster Dulles.[50]

It was in this arena that the Japanese trading companies – Mitsui, Mitsubishi and Sumitomo being among the largest – came into their own. These peculiarly Japanese concerns had their origins in the first period of that country's industrialization, when they were charged with providing the eyes and ears of those domestic corporations that ventured abroad. It was to them that fell the task of securing minerals for Japan during the post-war boom. They encountered especially fertile ground in South America, where they were helped along by the growth of a thriving Japanese–Brazilian trade in textiles, plus a healthy rate of Japanese emigration. Close ties were forged with Brazil, Peru and Chile, countries that were sources of the very materials needed by the furnaces at home. At the outset the trading companies focused on straightforward purchasing of the required ores and concentrates, but as time went on they were increasingly willing to negotiate long-term contracts that effectively underpinned new developments. Such support was crucial to the opening of the Brazilian and Australian iron ore mines. Their purchasing power, depth of knowledge and willingness to take long-term positions eventually gave them considerable influence in setting the price and overall competitive behaviour in many of the international mineral markets.[51]

One outcome of the southern hemisphere mineral developments was that ocean-going vessels carrying iron ore, copper concentrate, bauxite, alumina, steaming and coking coal and all manner of other minerals were daily docking at ports in Japan, Western Europe and the United States. Never before had the ships of the world been called on to move such a prodigious quantity of bulk materials across thousands of miles of ocean. It was an expensive business. Just how expensive it could be was brought forcefully home when Egypt's President Nasser nationalized the Suez Canal. The Arab–Israeli war that followed then closed it to traffic. Vessels that were sized to squeeze through the narrow Canal now had to be sailed around the southern tip of Africa. Freight rates soared, until at one point they sometimes equalled the cost of their cargo. The protracted crisis prompted a surge of shipbuilding and a re-evaluation of how oceanic commerce should be conducted.[52] For the first time there appeared a class of vessel called the bulk ore carrier. Soon there were some capable of carrying 50,000 tonnes, a far cry from the days when a clipper captain might dock at Swansea satisfied at having carried a cargo of 500 tonnes halfway around the world.

The post-war boom was an optimistic and prosperous time for the metallurgical industries of the world. The tendency to over-expansion was forgiven by the fact that demand for metals kept rising. Cartel behaviour diminished, although in Japan and Germany it remained a strong part of the business culture. This reduction in the strength of international cartels was in part due to post-war international agreements that frowned on such activity, and in part due to the fact that there was no real need for them. The success of the metals industries must be seen within its context. Virtually every sector of the broader economy was prospering, and lest it be thought that the mining and smelting companies occupied the heights in the same way as they had in the time of the Copper Barons, it is instructive to examine the rankings of the world's largest industrial enterprises for the year 1970.[53] Here we find that of the top 50 industrials, only four were metallurgical enterprises, compared to eight in 1912. And these four were all steelmakers. Indeed, all of the nine metallurgical companies that made it into the list of the top 100 industrials were closely connected to steel: US Steel (no. 14), LTV (16), Bethlehem Steel (31), British Steel (32), Thyssen (51), Yawata (67), National Coal Board (78), BHP (89) and Armco (97). The largest was still US Steel, but instead of being by far the largest industrial in the world, as it was in 1912, it had dropped to fourteenth. In part the decline in company rankings was due to industry fragmentation. During the boom, the concentrated, clubby atmosphere of the metals world had begun to dissolve as mineral discoveries made by small exploration outfits were converted into independent operating companies. But fragmentation was not the only explanation of the relative decline. It was also due to the reducing importance of metals in the industrialized economies.

STEEL: THE DECLINE OF A SYMBOL

The first generation of Soviet planners had never been in any doubt about the importance of metals to the national welfare. On the twelfth anniversary of the October Revolution of 1917, Stalin had declared: 'We are becoming a country of metals, an automobilised country, a tractorised country'.[54] In 1950, perhaps not too many statesmen would have said it in quite the same way, but Stalin's sentiment was widely shared. Without metals a country could not make its confident way in the world. Given the tasks ahead at that time the sentiment was justified. In 1975 it was not. To illustrate the transition let us examine the case of steel, for steel at the beginning of the post-war boom held in the political imagination a special status as both the reality and symbol of national power. It occupied a central place in the political discourse.

This was even the case in the United States where one may have thought that that country's share of one half of global steel production would have permitted the body politic to turn their attentions to other matters. Instead, the steelmakers drew the trenchant criticism of President Truman.[55] At the root of the problem was the question of the sustainability of the United States's global steel dominance. The steelmakers assumed that their post-war position could not last very long, and it was better to allow the countries of Europe some room to regain their lost ground. Their declared intention was to sit back awhile and allow peacetime demand to catch up with their new levels of production capacity. The attitude infuriated Truman, whose first concern was a strong and vibrant steel sector to underpin his country's superpower status. He threatened to build government-owned plants, and later sent in troops to occupy one of US Steel's mills when a looming strike drew only the most nonchalant of responses from the company's executives (the move was later ruled unconstitutional by the Supreme Court). The affair serves to underline the depth of feeling that the issue of steel could evoke in the immediate post-war years. The feelings were even stronger in the Soviet Union. Here the leadership and government planners maintained the direction that the nation had followed through the Stalinist era. The Five Year Plans were continued by Nikolai Krushchev, and steel was never far from the top of the priority list. The progress was impressive. In the single decade from 1950, production went up two and a half times.

The question of steel occupied no less a place in the considerations of European policy makers. Steel Plans, designed to encourage rebuilding, modernization and consolidation, were put in place in many countries. France launched its Plan of Modernization in 1947.[56] In Britain, the Labour government, exasperated at the slowness with which market forces were bringing about industry consolidation, decided to nationalize the fragmented industry. Although the Conservatives unwound the legislation two years later, even they eventually felt compelled, much against their instincts, to introduce the equivalent of a Steel Plan. When the Labour Party returned to power they created a government-owned entity, British Steel. In Germany, too, the industry became central in the debate about the country's economic future. On the larger canvas of Western Europe, steel came to be the vehicle for a much grander scheme.[57]

The idea of European unity had been around a long time without ever really making any concrete progress towards reality. Now the catastrophe of two wars had made the statesmen of Europe ready at last to look at a workable plan for achieving it. In 1950 Jean Monnet, a French economist and diplomat, presented just such a plan to Robert Schuman, the French foreign minister. In Monnet's blueprint, the first step towards European unification would be the creation of a common market for coal and steel. Why coal and steel? Underlying the plan was the widely held belief that the last 100 years of calamitous conflict between France and Germany had its origin, at least in part, in the race for each to dominate the other by sheer force of industrial might. If there existed a shared foundation of industrial strength, then this rivalry was deprived of sustenance on which to feed. The scheme would be called the European Coal and Steel Community. The governing institution was to be the High Authority, the practical tasks of which were to create a free trade zone in coal and steel, to encourage competition and, through subsidies and incentives, to assist the restructuring of the industry along more efficient lines.

Jumping ahead to 1975 we see a different picture. Thanks to mass production, steel was now in plentiful supply and was losing its grip in the halls of power. It was becoming just another commodity. To be sure, it could still mount a powerful emotional appeal, such as when the United States came into conflict with Japan over the growing success of Japanese steel exports. But this was more of an old-fashioned argument over industry protection than it was a matter of national strength and security. In Russia, too, the new reality was hitting home. Recognizing that the national focus on steel was in danger of becoming a monomania, in 1961 Krushchev publicly called for productive efforts and resources to be directed elsewhere. 'It is impossible', he said, 'to build communism by offering only machines and ferrous and non-ferrous metals. People must be able to eat and clothe themselves well, to have housing and other material and cultural conditions'.[58] We should not be deceived by the fact that Krushchev later reversed his comments when faced with an Old Guard reaction. He had correctly perceived that his country – once so desperately short of metals – would become glutted if production growth continued apace. In Europe a similar transition occurred. By 1975 the national Steel Plans had gone from managing growth to managing retrenchment. The European Coal and Steel Community was already largely forgotten. This is not to say it was a failure. Monnet was a political visionary, and his scheme had always appealed more to the statesman than to the industrialist. As a symbol of European unity, the European Coal and Steel Community had real power and served as the foundation stone for the European Economic Community. It had, however, little impact on conduct and competition between the steelmakers of Belgium, the Ruhr and northern France.

In only one industrialized country did steel hold a stronger grip on the national self-image than it had two-and-a-half decades before. That country was Japan. After the war, as we have seen, Japan had been placed under the baleful eye of General Douglas MacArthur. His mandate was to prevent the resurgence of Japanese aggression, and with this in mind he set about promoting Western-style political institutions and dismantling the power of the industrial–military elite who had led the country to war. Backed by legislation that went under the Orwellian title of the Law for the Elimination of Excessive Economic Concentration, he broke up the industrial groups, splitting the country's only large steelmaker, Japan Steel, into the two firms of Yawata and Fuji.[59] These two, along with a third independent company, NKK, dominated the fragmented industry. It was a time when government officials at the Ministry of International Trade and Industry (MITI) were surveying the economic landscape and formulating what were essentially rescue plans for different sectors of industry. In 1951 MITI delivered its First Rationalization Program for steel. The Ministry had no especially grand plans for the metal, and under the programme provisions the 'big three' were to be the main recipients of what governmental aid was available.

Into this cosy group burst Yataro Nishiyama, the head of Kawasaki Steel. This company had originally been part of the Kawasaki Heavy Industries group, and had become an independent entity during MacArthur's break up of the conglomerates. Nishiyama was not content with staying small. Seizing on the basic oxygen furnace technology recently proven in Europe, he proposed the construction at Chiba, just outside the Tokyo area, of a giant integrated steel works. The idea was strenuously opposed by government officials, and MITI refused it support. As arguments raged back and forth, Nishiyama proceeded anyway, and in the process set the trend for the industry. MITI soon recognized its error and began promoting the benefits of the Kawasaki approach, and especially of the basic oxygen furnace. The next fifteen years would see successively larger facilities constructed until Japan was home to an industry that could boast eight steel mills that were larger than the largest in the United States. The competitive dynamic that was created recalls the conditions in the United States at the end of the previous century. Behind a lofty tariff wall, Japanese producers cut prices savagely as each new domestic mill brought another great slug of steel on to the market. In turn the cheap steel made easier the growth of domestic car-making, shipbuilding and heavy machinery manufacture. Growing success as an exporter of these products added to the already thriving Japanese demand. Before long the output of each new mill was soaked up.

This competitive environment was too uncomfortable for the steel producers, and in time they sought the stability that lay in cooperation. So was created the system of *joshu-chosei*, or voluntary self-regulation. It comprised Yawata, Fuji and NKK as well as Kawasaki and two smaller producers, Sumitomo and Kobe. Under the skilful mediation of the head of Yawata Steel, this sextet would discuss their plans for expansion, sometimes agreeing to delay a project if this was for the good of the whole. Entitlement to new capacity was roughly based on historical sales, a fact that inevitably discriminated against the smaller firms with ambitions for greater things. It was a precarious unity. The crisis came when the industry attempted to arrange a concerted production cutback to help it through a sharp global economic recession. The proposal was that each mill should reduce output based on historical sales, and furthermore should delay expansion plans. As a fast-growing and export-oriented producer with plans for new facilities, Sumitomo refused to accept the plan. The company's president, Hyuga Hosai, expressed his frustration: 'There is no specific rule for the production reduction ... but always an ad-hoc way in which [Yawata and Fuji] try to control the latecomers'.[60]

It was called the 'Sumitomo rebellion', and provoked one of the rare occasions when MITI flexed its muscles. Unable to reach an agreement with the rebel, Yawata and Fuji asked MITI to intervene. Finding itself, too, rebuffed by Sumitomo, the Ministry threatened to cut off the company from its supply of imported coal. The Press responded by denouncing MITI's actions. Even with the hostility of its fellow cartel members and the government, it was probably only the difficult economic circumstances that forced Sumitomo eventually to pull back from its aggressive stance. The bitterness of the dispute meant the end for voluntary self-regulation. Everyone realized that some new scheme was needed, or there would be a return to the days of all-out competition. The answer was to create the equivalent of the proverbial 800-pound gorilla. The proposal to merge Yawata and Fuji met with the approval of the other members of *joshu-chosei*, and the deal was completed in 1970. The New Japan Steel Corporation, or Nippon Steel as it was known abroad, had become one of the largest steelmakers in the world. The faint rattling sound was Douglas MacArthur rolling in his grave.

INDEPENDENCIA

The post-war economic prosperity that buoyed along the wealthy nations was not shared in Latin America and Africa. This inequity and the feeling that their countries were being treated as little more than economic colonies prompted increasing numbers of people to challenge the existing order. National liberation movements of all shapes and hues gathered strength, frequently nourished by the struggle between the United States and the Soviet Union. Colonial empires were disintegrating. India had gained self-rule after the Second World War, and at the same time the peoples of the French colonies in Indo-China began their fight to throw out the European colonizers. The temper of the times was expressed by Harold Macmillan, the British Prime Minister, when he spoke of a wind of change blowing through Africa. When, in 1957, Kwame Nkrumah became the Prime Minister of an independent Ghana, the die was cast. By 1970 colonial Africa had vanished (except in white-ruled Rhodesia), leaving in its wake a host of infant nations.

Now that an increasing share of the world's mineral raw materials was being sourced from Africa and South America, the independence struggles brought turmoil to the world of mining. The mining companies had a high profile in these continents, and there was considerable political opposition to ownership of mineral resources by the large multinationals. Even in developed countries feelings could run high. In Canada the special assistant to the Prime Minister was moved to comment: 'It is not much of an exaggeration to say that Japanese governments have looked upon Canada in recent years as a large open-pit mine; as an endless and reliable source of raw materials to satisfy Japanese demands'.[61] In countries such as Brazil and Venezuela, where serious mineral development occurred only during the post-war boom, the host governments sought, often with success, to keep majority ownership in domestic hands. While the foreign companies might chafe under such restrictions, they learned to live with it. It was a different story in those countries where the mines had been developed and worked since well before the war. In these cases foreigners most often held control if not outright ownership. These arrangements came to be looked on with resentment and covetousness by the aspiring national leaders, and many of them were prepared to act upon their feelings. A wave of nationalizations swept through the industry. When it had run its course, the proportion of non-Communist global mine capacity that was held in government hands had risen from nearly zero to fully one-third.[62] Among the first companies to succumb was Union Minière.

When the Belgian flag was lowered for the last time in the Congolese capital of Kinshasa on 4 May 1960, it quickly became apparent how poorly the ground had been prepared for independence.[63] The new Prime Minister, Patrice Lumumba, insulted the gathering of Belgian dignitaries at the Independence Day celebrations. Days later the Force Publique, the feared and violent tool of Belgian repression, turned against their former masters as well as the civilian population. Within a week massacres had erupted across the country. Barely another week had passed when Moise Tsombe declared the southern province of Katanga to be an independent state. Old habits of paternalism died hard, and Tsombe's move was welcomed by the officers of Union Minière, disapproving as they were of the violent and radical regime that had installed itself in the north. They willingly allowed their taxation revenues to fill his coffers and pay for his mercenary army. It was a fatal mistake, for they had misunderstood the new geopolitical realities. In helping to restore order to the country, the United States backed not the business-friendly Republic of Katanga, whose uranium had helped them win the Second World War. Rather, they chose Joseph Mobutu, a military man who could keep the Congo safe from Communism. Tsombe was forced into exile. Mobutu remained faithful to the American cause, and he was not prepared to forgive the faithlessness of the one major corporation in his midst. The mines and refineries and all the other assets of Union Minière became the property of the Congolese government.

The Congo nationalization was one of many, and, apart from the massacres that accompanied the elimination of the Chinese from the tin-fields of Indonesia, it was surely conducted amid the most violence. But it was not the first. That honour belonged to the mountainous and desperately poor republic of Bolivia.[64] Bolivia was a single commodity economy. Tin, which had been found in large quantities underneath the silver in the mountain of Potosí, was the country's main export and the focal point of its politics. The industry had been dominated for decades by three absentee oligarchs, the most powerful of whom was Simon Patiño. They had done little for their country, exporting the profits from their mines to expand their smelting interests overseas and, so at least it appeared to their countrymen, to fund their lavish lifestyles. Resentment built until, in 1952, a successful coup brought to power a Left-wing political party, the MNR. Its most pressing policy challenge was not whether, but how to nationalize the mines. The new President, Victor Paz Estenssorro, knew that to proceed too quickly would invite the wrath of the Eisenhower administration. Accordingly he did what all politicians do when playing for time – he commissioned a study of the issue. He also sent his best man as Ambassador to the United States and set about assiduously cultivating an image of moderation and anti-Communism. When five months later the study team recommended in favour of nationalization with some compensation, it elicited little adverse comment from Washington. A similarly benign response greeted the peaceful process that unfolded just south of the Congo border. Here the newly created government of Zambia looked on the well-run and profitable copper operations of Anglo-American and the Selection Trust, and could not help but equate ownership of them with true independence. It created the Zambian Industrial and Mining Corporation, which duly acquired a majority holding in all the domestic copper operations.[65]

As befitted its long and influential role in international mining, it was Chile that provided the setting for the defining event of the nationalization movement. Let us go back to 1925. In that year the two copper giants Kennecott and Anaconda formed the Gran Minera. They bestrode the Chilean mining scene. Kennecott operated the old Braden mine of El Teniente and Anaconda ran Potrerillos and also Chuquicamata, which it had purchased for a colossal figure from the Guggenheims only later to find that that huge ore body held twice as much as anyone thought. The companies formed enclaves where platoons of American engineers managed everything from financial arrangements to the flotation mills. A Chilean who rose above the rank of shift boss was rare indeed. The marketing of the copper was controlled by the corporate sales office in New York. Royalties and taxes were paid by the companies on the basis of determinations made by their own accountants. In 1952 the Chilean Comptroller General could make a sworn statement to the Chilean Congress to the effect that no audit was ever undertaken to verify these numbers.[66] Consultation with the Chilean government was seen as a favour. World War II placed heavy demands on the mines, and the copper was paid for at prices unilaterally determined by the US government. This was accepted with good grace in Chile, in the expectation that their reward would come in later years. The recession of 1949 dashed those hopes. In that year a tariff on copper imports was under active consideration in the US Senate. As Senator Wallace Bennett of Utah put it: 'The policy of good neighbour should not be extended to the point of letting our own people go hungry.' When, during the Korean War, the US government decided to place a ceiling on the price of copper, Anaconda and Kennecott executives were called on to give advice and approve the level. No Chilean national was invited, even as an observer in the discussions.

In the face of all this, the policy makers in Santiago maintained for a long time an attitude of passive, if grudging, acceptance. This was due in part to their almost total ignorance of the milch cow in their midst. They dared not tamper for fear of doing it an injury, and their fears were deftly magnified by the company men. Yet the Korean War episode was one high-handed act too many, and perhaps also the spectacle of little Bolivia seizing control of its mines acted as an inspiration to the Chilean Congressmen. In 1952 they were emboldened to establish monopoly control over the sale of their country's copper. For good measure, taxes were raised until fully three-quarters of the profits from the Gran Minera were flowing to the Treasury coffers in Santiago. The tougher taxation regime was one thing, assigning to a handful of amateurs the task of selling 15 per cent of the world's copper was quite another. The sales monopoly was a disaster, and soon also the taxation policy was being blamed for a fall-off in the Chilean share of the global copper market. Soon opinion had swung the other way. Low taxation rates and generous incentives for investment were now declared the fit and proper manner in which to motivate properly the dedicated professionals who ran the big companies. When, five years later, overall production had not budged, disenchantment set in once more.

The success of Eduardo Frei in the Presidential elections of 1964 opened another chapter. He ran on a platform of the 'Chileanization' of copper. Perhaps mindful of the troubles in the Congo, executives of Kennecott now saw the writing on the wall. They approached Frei with a proposal to sell him a majority stake in their operations. It was a complicated deal, but one that added momentum to the President's programme. Then the old sales monopoly was re-established and taken a step further when Chile joined with Peru, Zambia and the Congo to form a copper exporters' cartel. Next Frei began to pressure Anaconda to follow the example of Kennecott and sell a majority share of its operations to the government. Unwilling to work with their long-time hosts, Anaconda unexpectedly offered to sell their entire stake in an arrangement that was sealed in July 1969. The Minister of Mines wept tears of joy as he announced the deal. Frei declared it to be Chile's Second Independence.

The elections of 1970 brought to power the Marxist Salvador Allende. Once in office, Allende proposed the final step – the total nationalization of Kennecott, Anaconda and the newcomer, Cerro Corporation. The action brought into being the Corporación del Cobre (Codelco), the world's largest copper company. Lest it be thought that Allende was acting against the wishes of the majority, the Nationalization Bill, when set before the Congress of Chile, was passed unanimously. It was probably the most popular act of a Presidency that would end in strikes, civil strife and, finally, a military coup. This last took place in September 1973, when General Augusto Pinochet led the armed forces that surrounded the Presidential palace, bombing it and eventually setting it on fire. The President did not leave the building alive.[67] Amid the campaign of murder and torture that followed the coup, one of the few of Allende's policies left intact was the creation of Codelco. One hundred and fifty years of foreign control of Chilean copper had finally come to an end. So, too, had the post-war boom, brought to a halt when delegates of the Organization of Petroleum Exporting Countries wrested control of the price of crude oil out of the hands of the industrialized nations and tripled it between September and December 1973.

Epilogue

Herbert Hoover died in 1964, the year of his 90th birthday. In the twilight of his life he may well have reflected on how his own career had paralleled the progress of the metallurgical industries in the twentieth century. As a mining engineer during the second industrial revolution, he had been in the thick of the struggles in which bountiful but troublesome ore bodies around the world were conquered and tamed. When the energy of that revolution had been spent, as a politician he played a central role in harnessing its fruits to the needs of the modern industrial state. Hoover's career is thus symbolic of the progress of metallurgical endeavour. For it is this interplay between forces technological and political (in the broad sense of the term) that has been characteristic of the development of mining and smelting throughout its long history. In this respect, of course, the industry is by no means unique. Similar forces, broadly consistent in strength and direction, have been at play in all industries. Indeed, the progress of mining and smelting has been more or less consistent with the broader rate and direction of industrial progress. It is not coincidence that mass-production techniques were introduced at Bingham Canyon at roughly the same time as Henry Ford was setting up his assembly plant at Dearborn, Illinois. Neither is it by accident that coal-based smelting came along soon after similar successful applications in glassmaking and the drying of hops for brewing – an observation that reminds us that many of the advances that have had a big impact on the industry had their genesis outside it.

Despite their similarities to the rest of the industrial world, mining and smelting have been subject to a particular set of influences that have given their evolution a flavour all its own. Their progress have been unusually susceptible to the influence of international political developments. This is because metals, being produced in only a relatively few locations yet demanded by all, have always been counted among that select group of goods that are traded in considerable quantities over long distances. Such a trade inevitably becomes caught up in the political, economic and social currents that are continually crossing the globe. Trade policies have been critical in shaping the industry. Chilean copper would have stayed in the ground without Huskisson's tariff reforms in post-Napoleonic Britain, and, one suspects, much of the Spanish colonial silver would have met the same fate had not the Chinese Emperor legalized the international commerce in silks and silver.

International political development in another of its guises – human migration – has also played a hugely important role in shaping the industry. This is because major mineral discoveries have very often come in the wake of migratory movements. Examples of this abound, perhaps the most spectacular being the story of Californian gold. In this case, as we have seen, the steady migration of American settlers into the environs of the Bay of San Francisco created a sharply increased demand for wood for construction of houses and suchlike. This in turn prompted John Sutter to build his famous sawmill along the banks of the American River on the wooded slopes of the Sierra Nevada. There is an additional way in which politics has driven progress in the industry. War and the preparation for it has been a constant driver of demand for metals, and for new metals. The twentieth century will suffice as an illustration. In a mere 50 years of violent post-industrial political restructuring, the demands of war created the three new mainstream metals of nickel, aluminium and uranium.

Let us dwell for a moment on this. It is a commonplace observation that military strength and the production of metals often go together. This has sometimes been interpreted to mean that the nation that possesses rich mineral deposits will possess also military might and, through it, international power. Examples come readily to mind – Spain in the sixteenth century, Sweden in the seventeenth, Great Britain in the nineteenth and the United States and the Soviet Union in the twentieth. But the observation misses the point. History shows us that it matters little what mineral wealth lies within a nation's borders or colonies. Of far more importance is the vigour with which any particular society goes about the task of extracting that wealth. The Stora Kopparberg held as much copper when Sweden was a moribund tributary state to Denmark as it did during the period of the Vasa kings. Russia's riches lay dormant when British iron and copper mines were leading the world, coming to life only with the ascent to power of the Bolsheviks. Germany twice challenged for global supremacy while possessing only a fraction of the mineral wealth of its adversaries.

Even more than the emergencies of war, the main driver of technological progress in mining and smelting has been the challenge of dealing with intractable ores. The sequence of events has become familiar as this history has unfolded. In a new mineral field, the ores are initially very rich and relatively easy to smelt to metal. As the mines go deeper and more workings are begun, they become much more difficult to treat. This prompts a crisis. The field may be abandoned, or, more usually, those who work it may simply become accustomed to poorer returns from their investment. However, with luck and persistence, the adversity can result in the development of a new metallurgical process. Liquation, amalgamation, flotation and cyanidation: all were born in this way. So too was the Thomas–Gilchrist process as well as many lesser advances and improvements. Only one other issue has come close as a driver of metallurgical advance, and that has been the struggle to secure adequate supplies of affordable fuel for the smelting process. To this we may attribute reverberatory smelting, the coke-fired blast furnace, the Cort process and, not least, the Bessemer process.

Metallurgical breakthroughs have often led to structural change in the industry. This is partly because a new metallurgical process generally results in a seismic shift in the economic viability of mineral regions. What was once considered waste rock is now a valuable and plentiful mineral source. The new process also generally results in the construction of larger works, which in turn brings to the fore an entirely new industry elite. The transformation of Potosí illustrates the case. On the great silver mountain, the water-driven machinery and carefully-constructed rock patios came at an expense that could only be justified by high daily throughputs. Overwhelmed by this flood of low-cost silver, the livelihoods of hundreds of Indian smelter-owners were destroyed.

The mechanism that drives metallurgical production towards larger plants and fewer owners has also driven the formation of cartels. Indeed, the history of metal production is liberally sprinkled with market-sharing agreements and price-fixing of all kinds. Superficially there are several sound reasons why these sorts of arrangements should be more prevalent in the metallurgical industries than elsewhere. For one thing metals, as finished products, are basic commodities with very little about them to distinguish one producer from another. Price, therefore, is the only competitive attribute. Second, major mineral deposits are usually host to several, often a multitude, of producers. The geographical proximity of these companies affords them common interests. As a result of these characteristics, so the reasoning goes, cooperation among mining and smelting firms is almost pre-ordained. Yet the history of mining and smelting tells us that actual competitive behaviour does not conform to this logic (except, perhaps, in the area of labour relations). Cornwall provides the classic example. America provides another, for during the Trust movement of the late nineteenth century, the metals sectors were amongst the last to be organized. There seems little to suggest that cartel behaviour in metals has been any more common or workable than in most other industries.

We have discussed the influence on the industry of difficult ores, of the search for cheap fuel as well as of war, human migration and other upheavals in the international political scene. But what of mineral discovery itself? It is the factor that frequently is accorded primacy as the driver of change in the industry. Yet the history reveals a different perspective. It suggests that it is more useful to view major mineral discoveries as the result of some other, more fundamental, development. In the case of central Europe, the Spanish colonies, the American West and southern Africa, the roots of discovery lay in the migration of alien settlers. In more recent times, the roots of discovery can often be traced to a policy decision to allow exploration on lands previously off-limits. Equally common is the case of the Western

Australian region of the Pilbara. Sixty years before its transformation into one of the world's richest iron ore provinces, Herbert Wootton, a government geologist, had declared it an 'iron country'. Nobody cared. Its location was far too distant and in terrain far too difficult to justify further enquiry. The passage of the years saw two crucial developments. Technological change came in the form of ocean-going bulk mineral freighters and, more importantly, political change saw the rise of a resource-hungry Japan. The lifting of the Australian embargo on the export of iron ore was the final step. Discovery of the Pilbara iron ore was just a matter of looking.

This history ends in the early 1970s – a finishing point that is somewhat arbitrary, but not entirely so. Those years saw the dawn of what appears to be a new era. The clearest point of demarcation between the post-war boom and this new era was the oil crisis of 1973, an event that coincided with, and in part caused, a general slowing of Western industrial growth. While slower growth, comparatively speaking, has been a feature of the new era, it is by no means its defining characteristic. It was in the mid-1970s that social and economic priorities that were in competition to industrial growth began to acquire a decisive influence in the politics of developed countries. The first tangible outcome of this was a series of legislative measures that set severe limits on the extent to which companies could dispose of their waste into the surrounding air and water. Miners and metals producers at first responded to the challenge with opposition, but realized eventually that they had neither majority opinion nor national necessity on their side. Management of waste and land disturbance has since become one of the drivers of technological change at all stages of mining and smelting. A second characteristic of the new era is the increasing importance of the developing nations in the consumption and production of metals. Of the eventual impact on the global industry of these and other developments it is still too early to tell. One thing can be said: they serve as a reminder that the path of progress for the metallurgical industries still has some turns to take.

References

PROLOGUE

1 Paul T. Craddock, *Early Metal Mining and Production* (Edinburgh, 1995), p. 123.
2 Robert Raymond, *Out of the Fiery Furnace* (London, 1986).
3 Raymond, *Fiery Furnace*, pp. 25–8.
4 R. F. Tylecote, *A History of Metallurgy* (London, 1992), p. 45.
5 Paul Einzig, *Primitive Money* (London, 1966), p. 203.
6 James Henry Breasted, *The Conquest of Civilisation* (New York, 1938), p. 205.
7 Raymond, *Fiery Furnace*, pp. 56–61.
8 Farley Mowat, *The Farfarers: Before the Norse* (Toronto, 1998), p. 67.
9 Robert Temple, *The Genius of China: 3,000 Years of Science, Discovery and Invention* (New York, 1986), p. 45.
10 Maurice Lombard, *Etudes d'Economie Medievale* (Paris, 1974), pp. 10–11.
11 J. Dutrizac, J. B. O'Reilly & R. J. MacDonald, 'Roman Lead Plumbing: Did it Really Contribute to the Fall of the Empire', *CIM Bulletin* (May 1982), pp. 111–14.
12 Robert Hartwell, 'A Cycle of Economic Change in Imperial China: Coal and Iron in Northeast China, 750–1350', *Journal of the Economic and Social History of the Orient*, X (1967), pp. 119–23.
13 William F. Collins, *Mineral Enterprise in China* (London, 1918), p. 37.
14 Jacques Gernet, *A History of Chinese Civilisation* (Cambridge, 1982), p. 415.
15 N. J. G. Pounds, *An Economic History of Medieval Europe* (London, 1974), p. 329; Gordon Tullock, 'Paper Money – A Cycle in Cathay', *Economic History Review*, IX/3 (1957), pp. 393–5.
16 Peter Spufford, *Money and its Uses in Medieval Europe* (London, 1988), p. 112.
17 Douglas Alan Fisher, *Steel from the Iron Age to the Space Age* (New York, 1967), pp. 23–4.
18 John V. Fine Jr, *The Late Medieval Balkans* (Ann Arbor, MI, 1987), pp. 199–201.

ONE · THE METALLURGICAL RENAISSANCE

1 Philippe Braunstein, 'Innovations in Mining and Metal Production in Europe in the Late Middle Ages', *Journal of European Economic History*,

XII (1983), pp. 586-9.
2 Philip Beitchman, *Alchemy of the Word: Cabala of the Renaissance* (New York, 1988).
3 Vannoccio Biringuccio, *Pirotechnia*, trans. C. S. Smith and M. T. Gnudi (New York, 1959).
4 Biringuccio, *Pirotechnia*, p. 41.
5 Biringuccio, *Pirotechnia*, p. 337.
6 Georg Agricola, *De Re Metallica*, trans. Herbert Hoover and Lou Hoover (London, 1912), p. 491.
7 Hermann Kellenbenz, *The Rise of the European Economy* (London, 1976), p. 109.
8 Danuta Molenda, 'Investments in Ore Mining in Poland from the 13th to the 17th Century', *Journal of European Economic History*, V (Spring 1976).
9 B. G. Awty, 'The Blast Furnace in the Renaissance Period: Haut Fourneau or Fonderie', *Newcomen Society*, LXI (1989–90), pp. 68–9.
10 T. K. Derry and Trevor Williams, *A Short History of Technology from the Earliest Times to A.D. 1900* (Oxford, 1960), pp. 137–40.
11 Marco Polo, *Le Devisement du Monde* (Geneva, 1998).
12 Jean Descola, *The Conquistadores* (New York, 1970), pp. 8–10.
13 Tony Nolan, *The Romantic World of Gold* (Sydney, 1980), pp. 30–31.
14 Cecil Jane, trans., *The Four Voyages of Christopher Columbus* (New York, 1988), p. 16.
15 Helmut Waszkis, *Mining in the Americas: Stories and History* (Cambridge, 1993), pp. 10–11.
16 William Weber Johnson, *Cortes* (Boston, MA, 1975).
17 Pierre Vilar, *A History of Gold and Money, 1450 to 1920* (London, 1976), pp. 103–11.
18 J. H. Clapham & E. Power, eds, *The Cambridge Economic History of the Middle Ages* (Cambridge, 1966), pp. 715–20.
19 Richard Ehrenberg, *Capital and Finance in the Age of the Renaissance: A Study of the Fuggers and their Connections* (London 1928), pp. 133–56.
20 Ehrenberg, *Capital and Finance*, p. 65.
21 Molenda, 'Investments in Ore Mining in Poland', p. 564.
22 Hermann Kellenbenz, ed., *Precious Metals in the Age of Expansion* (London, 1981), pp. 74–5.
23 *De Re Metallica*, p. 115.
24 Thomas Max Safley and Leonard N. Rosenband, *The Workplace before the Factory: Artisans and Proletarians, 1500–1800* (London, 1993), pp. 76–7.
25 Gerharde Benecke, *Maximilian I (1459–1519): An Analytical Biography* (London, 1982), p. 91.
26 Peter Blickle, *The Revolution of 1525: The German Peasant War from a New Perspective* (Baltimore, 1981), pp. 120–21.
27 Danuta Molenda, 'Technological Innovation in Central Europe between the XIVth and XVIIth Centuries', *Journal of European Economic History* (Fall 1995).
28 Ehrenberg, *Capital and Finance*, p. 142.

29 Peter J. Bakewell, 'Early Silver Mining in New Spain', in P. J. Bakewell, ed., *Mines of Silver and Gold in the Americas* (Aldershot, 1997), pp. 58–9.
30 Peter J. Bakewell, *Silver Mining and Society in Colonial Mexico, Zacatecas 1546–1700* (Cambridge, 1971), pp. 7–8.
31 Bartolome Arzans de Orsua y Vela, *Tales of Potosi*, trans. Francis Lopez-Morillas (Providence, RI, 1975).
32 *Tales of Potosi*, p. xiii.
33 J. U. Nef, 'Silver Production in Central Europe, 1450–1618', *Journal of Political Economy*, LIX (1941), p. 577; Vilar, *A History of Gold and Money*, p. 104.
34 Roland Turner and Stephen L. Goulden, eds, *Great Engineers and Pioneers in Technology, vol.* I (New York, 1981), pp. 175–7.
35 Ian Blanchard, 'English Lead and the International Bullion Crisis', in *Trade, Government and Economy in Pre-Industrial England* (London 1976), p. 25.
36 Jozek Vlachovic, 'Slovak Copper in the World Markets of the Sixteenth and First Quarter of the Seventeenth Centuries', p. 34.
37 M. B. Donald, *Elizabethan Copper: The History of the Company of the Mines Royal 1568–1605* (London, 1955), pp. 34–6.
38 Braunstein, 'Innovations in Mining and Metal Production', p. 583.
39 *De Re Metallica*, p. 95.
40 Alan Probert, 'Bartolome de Medina: The Patio Process and the Sixteenth Century Silver Crisis', *Mines of Silver and Gold in the Americas*, pp. 96–127. My following paragraphs draw heavily on Probert's essay.
41 A. J. Lynch, 'A History of Grinding: The World's First Technology', unpublished typescript.
42 Bakewell, *Silver Mining and Society in Colonial Mexico*, p. 154.
43 Bakewell, *Silver Mining and Society in Colonial Mexico*, p. 150.
44 *Tales of Potosi*, p. xxi.
45 Peter Bakewell, 'Technological Change in Potosi: The Silver Boom of the 1570s', *Mines of Silver and Gold in the Americas*, p. 59.
46 Peter Bakewell, *Miners of the Red Mountain: Indian Labour in Potosi, 1545–1650* (Albuquerque, NM, 1984).
47 Leslie Bethell, ed., *The Cambridge History of Latin America*, vol. II: *Colonial Latin America* (Cambridge, 1984), p. 130.
48 Arthur Preston Whitaker, *The Huancavelica Mercury Mine: A Contribution to the History of the Bourbon Renaissance in the Spanish Empire* (Cambridge, MA, 1941).
49 Bakewell, 'Technological Change in Potosi', pp. 91–3.
50 Jan Rogozinski, *A Brief History of the Caribbean* (New York, 1999), p. 40.
51 Earl J. Hamilton, 'American Treasure and the Rise of Capitalism (1500–1700)', *Economica* (November 1929), pp. 349–53.
52 K. N. Chaudhuri, *Trade and Civilisation in the Indian Ocean: An Economic History from the Rise of Islam to 1750* (Cambridge, 1985).
53 Albert Chan, *The Glory and Fall of the Ming Dynasty* (Norman, OK, 1982), pp. 131–3.

54 William F. Collins, *Mineral Enterprise in China* (London, 1918), pp. 27–8.
55 A.J.H. Latham and Heita Kawakatsu, *Japanese Industrialisation and the Asian Economy* (New York, 1994), pp. 73–5.
56 The Bureau of Mines of Japan, *Mining in Japan: Past and Present* (Tokyo, 1909), p. 20.
57 Edwin O. Reischauer and Albert M. Craig, *Japan: Tradition and Transformation* (Sydney, 1979), pp. 78–81.
58 Dennis Twitchett & Frederick W. Mote, eds, *The Cambridge History of China*, vol. VIII: *The Ming Dynasty, Pt. 2* (Cambridge, 1978), p. 398.
59 A. Kobata, 'The Production and Uses of Gold and Silver in 16th and 17th Century Japan', *Economic History Review*, XVIII (1965), p. 253.
60 'On the Track of the Manila Galleons', *National Geographic* (September 1990).
61 Albert Chan, 'Chinese–Philippine Relations in the Late Sixteenth Century to 1603', *Philippine Studies*, XXVI (1978), pp. 64–5.
62 Vlachovic, 'Slovak Copper', pp. 84–5.
63 Sven Rydberg, *The Great Copper Mountain: The Stora Story* (Hedemora, Sweden, 1988), p. 5.
64 Eli F. Heckscher, *An Economic History of Sweden* (Cambridge, MA, 1954), p. 72.
65 Vlachovic, 'Slovak Copper', p. 93.
66 Mag. Art. Kristoff Glamann, 'The Dutch East India Company's Trade in Japanese Copper, 1645–1736', *Scandinavian Economic History Review* (1953), pp. 52–3.
67 It is important to be reminded that the term 'voluminous' is a relative one. In its best years the weight of copper produced from Stora Kopparberg was probably little more than 3,000 tonnes. Globally, production may have hovered around 13,000 to 15,000 tonnes in a good year. Throughout the 1990s, global annual production was about 800 times higher.
68 Richard L. Garner, 'Long-term Silver Mining Trends in Spanish America: A Comparative Analysis of Peru and Mexico', *Mines of Silver and Gold in the Americas*, p. 227.

TWO · THE WATT ENGINE

1 M. B. Donald, *Elizabethan Monopolies: The History of the Company of Mineral and Battery Works, 1565–1604* (Edinburgh, 1961).
2 M. B. Donald, *Elizabethan Copper: The History of the Company of Mines Royal, 1568–1605* (London, 1955).
3 Donald, *Elizabethan Copper*, p. 143.
4 E. J. Hobsbawm, *The Age of Revolution: Europe 1789–1848* (London, 1962), pp. 33–7.
5 G. J. Hollinger-Short, 'Gunpowder and Mining in 16th- and 17th-century Europe', *History of Technology* (1985), p. 32.
6 Francis Pierre, *Mines de Le Thillot (Vosges)* (Paris, 1993), p. 100.

7 G. J. Hollinger-Short, 'The Use of Gunpowder in Mining: A Document of 1627', *History of Technology* (1983), pp. 113–14.
8 J. U. Nef, 'An Early Energy Crisis and its Consequences', *Scientific American* (November 1977), pp. 140–51.
9 G. Hammersley, 'The State and the English Iron Industry in the Sixteenth and Seventeenth Centuries', *Trade, Government and Economy in Pre-industrial England* (London, 1976), p. 168.
10 Eleanor S. Godfrey, *The Development of English Glassmaking, 1560–1640* (Oxford, 1975), pp. 59–65.
11 J. R. Harris, *The Copper King: A Biography of Thomas Williams of Llanidan* (Liverpool, 1964), p. 3.
12 Joan Day & R. F. Tylecote, eds, *The Industrial Revolution in Metals* (London, 1991).
13 D. B. Barton, *Essays in Cornish Mining History*, vol. II (Truro, 1968), p. 85.
14 J. R. Harris, *The British Iron and Steel Industry, 1700–1850* (Basingstoke, 1988), p. 31.
15 H. R. Schubert, 'Abraham Darby and the Beginnings of the Coke–Iron Industry', *Journal of the Iron and Steel Institute* (September 1959), p. 3.
16 D. B. Barton, *The Cornish Beam Engine* (Truro, 1965), pp. 7–9.
17 Howard Jones, *Steam Engines: An International History* (London, 1973), p. 20.
18 Ivor B. Hart, *James Watt and the History of Steam Power* (London, 1958), p. 135.
19 H. W. Dickinson, *Matthew Boulton* (Cambridge, 1937), p. 72.
20 T. H. Marshall, *James Watt* (London, 1925), p. 116.
21 Marshall, *James Watt*, p. 120.
22 A. L. Prowse, *Tudor Cornwall: Portrait of a Society* (London, 1941), p. 62.
23 George Randall Lewis, *The Stannaries: A Study of the English Tin Miner* (Cambridge, MA, 1924).
24 Henry Hamilton, *The English Brass and Copper Industries to 1800* (London, 1967), p. 159.
25 Day & Tylecote, *The Industrial Revolution in Metals*, p. 110.
26 Richard Chadwick, 'Copper: The British Contribution', *CIM Bulletin* (October 1983), p. 85.
27 Robert R. Toomey, *Vivian and Sons, 1809–1924: A Study of the Firm in the Copper and Related Industries* (New York, 1985), pp. 175–7.
28 Hamilton, *English Brass and Copper Industries*, p. 280.
29 J. R. Harris, 'Copper and Shipping in the Eighteenth Century', *Economic History Review*, XIX (1966), pp. 550–68.
30 J. B. Richardson, *Metal Mining* (London, 1974), p. 93.
31 Harris, *The Copper King*, p. 35.
32 J. R. Harris & R. O. Roberts, 'Eighteenth-century Monopoly: The Cornish Metal Company Agreements of 1785', *Business History*, I (1961), pp. 69–82.
33 Harris, *The Copper King*, p. 61.
34 Harris, *The Copper King*, p. 87.
35 M. S. Anderson, *Europe in the Eighteenth Century, 1713–1783* (London,

1961), p. 99.

36 This well-known oil painting is in the Science Museum, London. It is reproduced in, for example, William Vaughan's widely available *Romantic Art* (London, 1978), p. 22, fig. 10.

37 Referred to in 'And did those feet in ancient time ...', the introductory poem to *Milton: A Poem in Two Books* (1804); see *The Complete Poetry & Prose of William Blake*, edited by David V. Erdman (New York, 1982), p. 95.

38 Preceptor K.-G. Hildebrand, 'Foreign Markets for Swedish Iron in the 18th Century', *Scandinavian History Review*, IV (1958), p. 9.

39 Thomas Southcliffe Ashton, *Iron and Steel in the Industrial Revolution* (Manchester, 1963), pp. 91–3.

40 J. L. Hammond & Barbara Hammond, *The Rise of Modern Industry* (London, 1966), p. 139.

41 Alan Birch, *The Economic History of the British Iron and Steel Industry* (London, 1947), p. 47.

42 Eli F. Heckscher, *The Continental System: An Economic Interpretation* (Gloucester, MA, 1964), pp. 262–4.

43 John H. Coatsworth, 'The Mexican Mining Industry in the Eighteenth Century', *Mines of Silver and Gold in the Americas* (Aldershot, 1997), pp. 263–4.

44 John Fisher, 'Silver Production in the Viceroyalty of Peru, 1776–1824', *Mines of Silver and Gold in the Americas*, p. 296.

45 J. Fred Rippy, *British Investments in Latin America, 1822–1949* (Minneapolis, 1959), pp. 17–21.

46 W. E. Minchinton, *The British Tinplate Industry: A History* (Oxford, 1957), p. 12.

47 Llewellyn Woodward, *The Age of Reform* (Oxford, 1962), pp. 71–2.

48 Simon Collier & William F. Sater, *A History of Chile, 1808–1994* (Cambridge, 1996), pp. 76–8.

49 John Mayo, *British Merchants and Chilean Development: 1851–1886* (Boulder, CO, 1987), pp. 20–23.

50 Thomas Ashton & Joseph Sykes, *The Coal Industry of the Eighteenth Century* (Manchester, 1929), p. 43.

51 D. B. Barton, *A History of Tin Mining and Smelting in Cornwall* (Exeter, 1989), p. 88.

52 Ashton & Sykes, *Coal Industry of the Eighteenth Century*, p. 41.

53 Woodward, *Age of Reform*, p. 151.

54 A. J. Wilson, *The Pick and the Pen* (London, 1979).

55 Woodward, *Age of Reform*, p. 156.

56 Roger Burt, *John Taylor: Mining Entrepreneur and Engineer* (Buxton, 1977), p. 15.

57 Roger Burt, ed., *Cornish Mining: Essays on the Organization of Cornish Mines and the Cornish Mining Economy* (Newton Abbot, 1969), pp. 9–12.

58 Charles Harvey & Peter Taylor, 'Mineral Wealth and Economic Development: Foreign Direct Investment in Spain, 1851–1913', *Economic History Review*, II (1987), p. 189.

59 Roger Burt, 'The London Mining Exchange 1850–1900', *Business History*, XIV (1972), p. 124.
60 Henry Faul & Carol Faul, *It Began with a Stone: A History of Geology from the Stone Age to the Age of Plate Tectonics* (New York, 1983), p. 94.
61 Edward Bailey, *Geological Survey of Great Britain* (London, 1952), pp. 11–13.
62 Faul & Faul, *It Began with a Stone*, pp. 112–14.
63 D. B. Barton, *A History of Copper Mining in Cornwall and Devon* (Truro, 1968), p. 34.
64 H. J. Habbakuk & M. Postan, eds, *The Cambridge Economic History of Europe*, vol. VI (Cambridge, 1965), p. 227.
65 W. T. Jackman, *The Development of Transportation in Modern England* (London, 1966), p. 481.
66 Philip S. Bagwell, *The Transport Revolution from 1770* (London, 1974), p. 92.
67 J. C. Carr & W. Taplin, *History of the British Steel Industry* (Cambridge, MA, 1962), p. 20.
68 Carr & Taplin, *History of the British Steel Industry*, p. 19.
69 Cort's puddling relied on the same principle as Bessemer's air-blowing. Both brought oxygen into contact with the molten sulphur. The difference between the two was that Bessemer's technique was far more rapid. It also had the effect of removing much more of the carbon left over from the coking coal in the blast furnace. It is in its lower level of carbon by which steel is distinguished from iron.
70 W. M. Williams, 'Manganese and its Importance to Nineteenth-century Metallurgy', *CIM Bulletin* (April 1982), p. 159.
71 Carr & Taplin, *History of the British Steel Industry*, p. 27.
72 Leslie Bethell, ed., *The Cambridge History of Latin America*, vol. III (Cambridge, 1984), p. 333; Collier & Sater, *History of Chile*, p. 84.
73 Maurice Zeitlin, *The Civil Wars in Chile* (Princeton, NJ, 1984).
74 For this to be workable, rapid communications were required, a need that was satisfied when the first transatlantic telegraph cable was laid in 1866.
75 J. Kennedy, *A History of Malaya* (New York, 1970), p. 131.
76 Wong Lin Ken, *The Malayan Tin Industry to 1914* (Tucson, AZ, 1965), p. 24.
77 Barton, *History of Tin Mining and Smelting in Cornwall*, p. 151.

THREE · NEW FRONTIERS

1 Victor S. Clark, *History of Manufactures in the United States*, vol. I: *1607–1860* (New York, 1929), pp. 220–40.
2 Frederic L. Paxson, *History of the American Frontier, 1763–1893* (Boston, MA, 1924), p. 323.
3 T. A. Rickard, *A History of American Mining* (New York, 1932), pp. 155–60.
4 William Issel & Robert W. Cherry, *San Francisco, 1865–1932: Politics,*

Power and Urban Development (Berkeley, CA, 1986), p. 7.
5 Rickard, *History of American Mining*, p. 19.
6 Henry Faul & Carol Faul, *It Began with a Stone: A History of Geology from the Stone Age to the Age of Plate Tectonics* (New York, 1983), p. 174.
7 Geoffrey Blainey, *The Rush that Never Ended: A History of Australian Mining* (Melbourne, 1963), pp. 106–8.
8 Nicholas Makeev & Valentine O'Hara, *Russia* (London, 1925), pp. 32–3.
9 Hugh D. Hudson, *The Rise of the Demidov Family and the Russian Iron Industry in the Eighteenth Century* (Newtonville, MA, 1986), pp. 35–42.
10 Ian Blanchard, *Russia's Age of Silver: Precious Metal Production and Economic Growth in the Eighteenth Century* (London, 1989).
11 W. P. Morrell, *The Gold Rushes* (London, 1940), p. 48.
12 Thomas D. Clark, *Frontier America: The Story of the Westward Movement* (New York, 1959), p. 523.
13 Michael S. Durham, *Desert Between the Mountains: Mormons, Miners, Padres, Mountain Men and the Opening of the Great Basin, 1772–1869* (New York, 1997).
14 Paxson, *History of the American Frontier*, p. 372.
15 Rickard, *History of American Mining*, p. 24.
16 Rickard, *History of American Mining*, p. 28.
17 Oscar Lewis, *San Francisco: Mission to Metropolis* (San Diego, CA, 1980), p. 50.
18 Lewis, *San Francisco*, p. 52.
19 J. S. Holliday, *The World Rushed In: The California Gold Rush Experience* (New York, 1981), p. 297.
20 Douglas Fetherling, *The Gold Crusades: A Social History of Gold Rushes, 1849–1929* (Toronto, 1997), p. 37.
21 Fetherling, *The Gold Crusades*, p. 27.
22 Rickard, *History of American Mining*, p. 33.
23 John Walton Caughey, *The California Gold Rush* (Berkeley, CA, 1948), p. 228.
24 Charles Ross Parke, *Dreams to Dust: A Diary of the California Gold Rush, 1849–1850* (Lincoln, NE, 1989), entry for 18 September 1849.
25 Holliday, *The World Rushed In*, p. 320.
26 Ralph Mann, *After the Gold Rush: Society in Grass Valley and Nevada City, California, 1849–1870* (Stanford, CA, 1982).
27 David Lavender, *California: A Bicentennial History* (New York, 1976), pp. 87–92.
28 James J. Rawls & Richard J. Orsi, eds, *A Golden State: Mining and Economic Development in Gold Rush California* (Berkeley, CA, 1999), pp. 168–9.
29 Jan Kociumbas, *The Oxford History of Australia*, vol. II (Oxford, 1992), p. 297.
30 Blainey, *The Rush that Never Ended*, p. 14.
31 Peter Burroughs, *Britain and Australia 1831–1855: A Study in Imperial Relations and Crown Lands Administration* (Oxford, 1967), Appendix II.
32 J. Holland Rose, A. P. Newton & E. A. Benians, *The Cambridge History*

of the British Empire, vol. II (Cambridge, 1940), pp. 429–30.
33 Blainey, *The Rush that Never Ended*, p. 8.
34 Kociumbas, *Oxford History of Australia*, p. 300.
35 Geoffrey Serle, *The Golden Age: A History of the Colony of Victoria, 1851–1861* (Melbourne, 1963), p. 9.
36 Serle, *The Golden Age*, p. 23.
37 See Geoffrey Blainey, *The Tyranny of Distance: How Distance Shaped Australia's History* (Melbourne, 1968).
38 Blainey, *The Rush that Never Ended*, p. 21.
39 Frank Cusack, *Bendigo: A History* (Melbourne, 1973), p. 117.
40 W. E. Adcock, *The Gold Rushes of the Fifties* (Melbourne, 1977), pp. 43–5.
41 Alexander Sutherland, *Victoria and its Metropolis* (Melbourne, 1888), pp. 407–10.
42 T. E. Gregory, 'Gold Standard', *The Times* [London] (20 June 1933).
43 Pierre Vilar, *A History of Gold and Money, 1450 to 1920* (London, 1976), pp. 322–3.
44 William Wiseley, *A Tool of Power: The Political History of Money* (New York, 1977), pp. 48–9.
45 Clyde A. Milner, Carol O. O'Connor & Martha A. Sandweiss, *The Oxford History of the American West* (Oxford, 1994), pp. 202–6.
46 Elliott West, *The Contested Plains: Indians, Goldseekers and the Rush to Colorado* (Lawrence, KS, 1998), pp. 303–5.
47 Blainey, *The Rush that Never Ended*, p. 83.
48 Eliot Lord, *Comstock Mining and Miners* (San Diego, CA, 1959), p. 38.
49 Theo F. Van Wagenen, *International Mining Law* (London, 1918), p. 100.
50 A. H. Ricketts, *American Mining Law, with Forms and Precedents* (Sacramento, CA, 1943), p. xxvii.
51 Lord, *Comstock Mining and Miners*, p. 393.
52 Irving Stone, *Men to Match my Mountains: The Opening of the Far West, 1840–1900* (London, 1946), p. 260.
53 Stone, *Men to Match my Mountains*, pp. 256–8.
54 Dan de Quille, *The Big Bonanza* (New York, 1947), pp. 174–7.
55 Lord, *Comstock Mining and Miners*, p. 354.
56 Lord, *Comstock Mining and Miners*, pp. 311–12.
57 Arthur Birnie, *An Economic History of Europe, 1760–1939* (London, 1948), p. 81.
58 John F. Stover, *American Railroads* (Chicago, 1997), p. 64.
59 Glenn Chesney Quiett, *Pay Dirt: A Panorama of American Gold-Rushes* (New York, 1936), p. 263.
60 Carroll Fenton & Mildred Fenton, *Giants of Geology* (New York, 1952), p. 236.
61 Fenton & Fenton, *Giants of Geology*, pp. 239–40.
62 Dee Brown, *Bury my Heart at Wounded Knee: An Indian History of the American West* (New York, 1970).
63 Roger Thompson, *The Golden Door: A History of the United States of America* (London, 1969), p. 313.

FOUR · THE COPPER BARONS

1 David Avery, *Not on Queen Victoria's Birthday* (London, 1974).
2 Fred Aftalion, *A History of the International Chemical Industry* (Philadelphia, 1991), p. 12.
3 S. G. Checkland, *The Mines of Tharsis: Roman, French and British Enterprise in Spain* (London, 1967), p. 91.
4 Checkland, *The Mines of Tharsis*, pp. 118–20.
5 Charles E. Harvey, *The Rio Tinto Company: An Economic History of a Leading International Mining Concern, 1873–1954* (Penzance, 1981), pp. 6–11.
6 Ibid, p. 30.
7 William B. Gates, *Michigan Copper and Boston Dollars: An Economic History of the Michigan Copper Mining Industry* (New York, 1951), pp. 12–13.
8 Geoffrey Blainey, 'A Theory of Mineral Discovery: Australia in the Nineteenth Century', *Economic History Review*, XXVI (1973), pp. 298–313.
9 T. A. Rickard, *A History of American Mining* (New York, 1933), p. 239.
10 Louis Agassiz, *Studies on Glaciers* (New York, 1937).
11 F. E. Richter, 'The Copper Mining Industry in the United States, 1845–1925', *Quarterly Journal of Economics*, XLI (1927), p. 250.
12 Michael P. Malone, *The Battle for Butte: Mining and Politics on the Northern Frontier 1864–1906* (Seattle, 1981), p. 28.
13 William Bronson and T. H. Watkins, *Homestake: The Centennial History of America's Greatest Gold Mine* (San Francisco, 1977), p. 27.
14 Malone, *The Battle for Butte*, p. 29.
15 Robert Glass Cleland, *A History of Phelps Dodge 1834–1950* (New York, 1952), pp. 100–103.
16 Thomas R. Navin, *Copper Mining and Management* (Tucson, AZ, 1978), p. 114.
17 F.E.B. Andrews, 'The Late Copper Syndicate', *Quarterly Journal of Economics*, III (1889), p. 507.
18 Nicol Brown & Charles C. Turnbull, *A Century of Copper* (London, 1900); *Engineering and Mining Journal* (22 April 1893), p. 370.
19 Joseph Newton, *Metallurgy of Copper* (New York, 1942), p. 484.
20 Michael White, *Acid Tongues and Tranquil Dreamers: Tales of Bitter Rivalry that Fuelled the Advance of Science and Technology* (New York, 2001).
21 Ronald W. Clark, *Edison: The Man who Made the Future* (London, 1977); Edward Tatnall Canby, *A History of Electricity* (New York, 1963), p. 92.
22 Navin, *Copper Mining and Management*, p. 65.
23 Thomas Greaves & William Culver, eds, *Miners and Mining in the Americas* (Manchester, 1985), p. 74.
24 Harold Blakemore, *British Nitrates and Chilean Politics, 1886–1896* (London, 1974), pp. 114–15.
25 Christopher J. Schmitz, *World Non-Ferrous Metal Production and Prices, 1700–1976* (London, 1979).

26 Alfred D. Chandler, *Scale and Scope: The Dynamics of Industrial Capitalism* (Cambridge, MA, 1990).
27 Erik Bergengren, *Alfred Nobel: The Man and his Work* (London, 1962), p. 44.
28 Martha Moore Trescott, *The Rise of the American Electrochemicals Industry, 1880–1910* (Westport, CT, 1981), p. 61.
29 Joseph G. Rayback, *A History of American Labour* (New York, 1959), pp. 195–6.
30 John Moody, *The Truth about Trusts* (New York, 1968), pp. 454–67.
31 Kenneth Warren, *Triumphant Capitalism: Henry Clay Frick and the Industrial Transformation of America* (Pittsburgh, 1996).
32 Lewis Corey, *The House of Morgan: A Social Biography of the Masters of Money* (New York, 1930), pp. 262–7.
33 H. H. Langton, *James Douglas: A Memoir* (Toronto, 1940), p. 50.
34 B. Webster Smith, *The World's Great Copper Mines* (London, 1967), p. 29.
35 Jamie Swift, ed., *The Big Nickel: Inco at Home and Abroad* (Kitchener, Ontario, 1977), p. 19.
36 John Deverell, ed., *Falconbridge: Portrait of a Canadian Multinational* (Toronto, 1975), p. 22.
37 Swift, *The Big Nickel*, p. 21.
38 Isaac Marcosson, *Anaconda* (New York, 1957), pp. 93–5.
39 C. B. Glasscock, *The War of the Copper Kings: Builders of Butte and Wolves of Wall Street* (Indianapolis, 1935), pp. 245–7.
40 John H. Davis, *The Guggenheims: An American Epic* (New York, 1978), pp. 58–60.
41 Irving Stone, *Men to Match my Mountains: The Opening of the Far West, 1840–1900* (New York, 1946).
42 Davis, *The Guggenheims*, p. 61.
43 Marvin D. Bernstein, *The Mexican Mining Industry, 1890–1950* (Albany, NY, 1965), p. 38.
44 James E. Pell, *Ores to Metals: The Rocky Mountain Smelting Industry* (Lincoln, NE, 1979), p. 236.
45 James Howard Bridge, *The Inside Story of the Carnegie Steel Company* (New York, 1903), p. 259.
46 David A. Walker, *Iron Frontier: The Discovery and Early Development of Minnesota's Three Ranges* (St Paul, MN, 1979), p. 133.
47 Joseph Frazier Wall, *Andrew Carnegie* (New York, 1959), p. 596.
48 Matthew Josephson, *The Robber Barons: The Great American Capitalists* (London, 1962), p. 390.
49 Crowell & Murray, *Inco The Iron Ores of Lake Superior* (Cleveland, OH, 1917), pp. 48–9.
50 A. B. Parsons, *The Porphyry Coppers* (New York, 1933), pp. 61–2.
51 W. H. Dennis, *A Hundred Years of Metallurgy* (Chicago, 1963), p. 41.
52 Geoffrey Blainey, *The Peaks of Lyell* (Melbourne, 1967).
53 Geoffrey Blainey, *The Rise of Broken Hill* (Melbourne, 1968), p. 25.
54 E. Cocks and B. Walters, *A History of the Zinc Smelting Industry* (London, 1968), p. 7.

55 Rickard, *History of American Mining*, p. 398.
56 D. W. Fuerstenau, ed., *Froth Flotation: 50th Anniversary Volume* (New York, 1962), p. 42.
57 Dennis, *Hundred Years of Metallurgy*, p. 38.
58 Fuerstenau, *Froth Flotation*, p. 21.
59 Robert H. Ramsay, *Men and Mines of Newmont: A Fifty–Year History* (New York, 1973), p. 21.
60 Ira Joralemon, *Copper: The Encompassing Story of Mankind's First Metal* (Berkeley, CA, 1973), pp. 255–9.
61 Harvey O'Connor, *The Guggenheims: The Making of an American Dynasty* (New York, 1937), pp. 268–74.
62 Christopher J. Schmitz, 'The World's Largest Industrial Companies of 1912', *Business History*, XXXVII/4 (1995), pp. 85–96.

FIVE · THE GOLD FACTORIES

1 Winston S. Churchill, *A History of the English Speaking Peoples*, vol. III (London, 1957), p. 279.
2 Marq de Villiers, *White Tribe Dreaming: Apartheid's Bitter Roots* (Toronto, 1987), pp. 122–8.
3 C. W. de Kiewiet, *A History of South Africa, Social and Economic* (Oxford, 1941).
4 Brian Roberts, *The Diamond Magnates* (London, 1988), pp. 1–2.
5 Roberts, *Diamond Magnates*, p. 10.
6 Godehard Lenzen, *The History of Diamond Production and the Diamond Trade* (London, 1970), p. 149.
7 Richard Lewinsohn, *Barney Barnato: from Whitechapel Clown to Diamond King* (London, 1937), pp. 77–8.
8 Colin Newbury, *The Diamond Ring: Business, Politics and Precious Stones in South Africa, 1867–1947* (Oxford, 1989), p. 45.
9 Roberts, *Diamond Magnates*, p. 172.
10 Robert I. Rotberg, *The Founder: Cecil Rhodes and the Pursuit of Power* (Oxford, 1988), pp. 55–9.
11 Paul Emden, *Randlords* (London, 1935), p. 66.
12 Lenzen, *Diamond Production and the Diamond Trade*, p. 157.
13 Niall Ferguson, *The House of Rothschild: The World's Banker, 1849–1998* (London, 1998), pp. 297–304.
14 Ferguson, *House of Rothschild*, p. 353.
15 Emden, *Randlords*, pp. 94–6.
16 Emden, *Randlords*, pp. 98–101.
17 Geoffrey Wheatcroft, *The Randlords* (London, 1985), p. 90.
18 Wheatcroft, *Randlords*, pp. 68–9.
19 Owen Letcher, *The Gold Mines of Southern Africa: The History, Technology and Statistics of the Gold Industry* (Johannesburg, 1936), p. 96.
20 A. P. Cartwright, *The Gold Miners* (Cape Town, 1962), p. 95.
21 F. Addington Symonds, *The Johannesburg Story* (London, 1953), p. 99.
22 S. G. Checkland, *The Mines of Tharsis: Roman, French and British*

Enterprise in Spain (London, 1967), p. 127.
23 Fathi Habashi, 'One Hundred Years of Cyanidation', *CIM Bulletin* (September 1987), p. 110.
24 A. L. Lougheed, 'The Cyanide Process and Gold Extraction in Australia, 1888–1913', Working Paper no. 51 (June 1985).
25 Dan Fivehouse, *The Diamond Drilling Industry* (London, 1976), p. 120.
26 Robert Vicat Turrell & Jean-Jacques Van Helten, 'The Rothschilds, the Exploration Company and Mining Finance', *Business History*, XXVIII (1986), p. 187.
27 John Hays Hammond, *The Autobiography of John Hays Hammond* (New York, 1935).
28 M. E. Chamberlain, *The Scramble for Africa* (London, 1999).
29 Antony Thomas, *Rhodes* (London, 1996).
30 S. E. Katzenellenbogen, *Railways and the Copper Mines of Katanga* (Oxford, 1973), p. 11.
31 George Martelli, *Leopold to Lumumba: A History of the Belgian Congo, 1877–1960* (London, 1962), p. 192.
32 Elizabeth Pakenham, *Jameson's Raid* (London, 1960), p. 46.
33 Geoffrey Blainey, 'The Lost Causes of the Jameson Raid', *Economic History Review*, XVIII (1965), pp. 350–66.
34 D. Jacobsson, *Fifty Golden Years of the Rand, 1886–1936* (London, 1937), p. 77.
35 G. H. Le May, *The Afrikaaners: An Historical Interpretation* (Oxford, 1995), pp. 103–10.
36 Robin W. Winks, ed., *British Imperialism: Gold, God, Glory* (New York, 1963), p. 22.
37 Brian Kennedy, *A Tale of Two Mining Cities: Johannesburg and Broken Hill 1885–1925* (Melbourne, 1984).
38 A. B. Parsons ed., *Seventy-five Years of Progress in the Mineral Industry, 1871–1946* (New York, 1947), p. 377.
39 A. J. Wilson, *The Professionals: The Institution of Mining and Metallurgy, 1892–1992* (London, 1992), p. 18.
40 Parson, *Seventy-five Years of Progress*, p. 380.
41 Such a perception would die hard. Herbert Hoover told the tale of a cruise from New York to London during which he had made the acquaintance of a lady of obviously considerable social standing. On learning that the future US President was content to describe himself as a mining engineer, she exclaimed 'Oh, I thought you were a gentleman!'
42 Charles Harvey & Jon Press, 'Overseas Investment and the Professional Advance of British Metal Miners, 1851–1914', *Economic History Review*, XLII (1989), p. 72.
43 Eric Rosenthal, *Gold! Gold! Gold!: The Johannesburg Gold Rush* (Johannesburg, 1970), p. 247.
44 R. G. Hartley, *The 1904 Watershed in Bewick Moreing's Western Australian Goldmining Activities* (Perth, 1992), p. 48.
45 Hartley, *The 1904 Watershed*, p. 61.
46 Pierre Berton, *Klondike: The Last Great Gold Rush, 1896–1899* (Toronto,

1972), p. 40.

47 Morris Zaslow, *The Opening of the Canadian North, 1870–1914* (Toronto, 1971), p. 110.

48 Morgan B. Sherwood, ed., *Alaska and its History* (Seattle, 1967), p. 380.

49 Clark C. Spence, *The Northern Gold Fleet: Twentieth-Century Gold Dredging in Alaska* (Chicago, 1996), p. 9.

50 Lewis Green, *The Gold Hustlers* (Anchorage, 1977), p. 29.

51 A. Ascain and J.-M. Arnaud, *History of Money and Finance* (London, 1967), p. 67.

52 John Chown, *A History of Money From AD 800* (London, 1994).

53 Giulio M. Gallarotti, *The Anatomy of an International Monetary Regime: The Classical Gold Standard, 1880–1914* (Oxford, 1995), p. 151.

54 Milton Friedman, *Money Mischief: Episodes in Monetary History* (New York, 1992), pp. 51–80.

55 David B. Anderson, *William Jennings Bryan* (Boston, 1981).

56 Christopher J. Schmitz, *World Non-Ferrous Metal Production and Prices, 1700–1976* (London, 1979), pp. 81–5.

57 John V. Dorr, *Cyanidation and Concentration of Gold and Silver Ores* (New York, 1936), pp. 4–5.

58 *Engineering and Mining Journal* (6 August 1892), p. 121.

59 Rosenthal, *Gold! Gold! Gold!: The Johannesburg Gold Rush*, p. 219.

60 Wheatcroft, *The Randlords*, p. 222.

61 David Yudelman, *The Emergence of Modern South Africa* (New York, 1983), p. 193.

62 A. P. Cartwright, *The Gold Miners* (Cape Town, 1962), p. 166.

63 Labour History Group, *The 1922 White Mineworker's Strike* (Salt River, S.A., 1986), p. 30.

64 Jacobsson, *Fifty Golden Years of the Rand*, p. 92.

65 Harvey O'Connor, *The Guggenheims: The Making of an American Dynasty* (New York, 1937), p. 317.

66 James W. Byrkit, *Forging the Copper Collar: Arizona's Labor-Management War of 1901–1921* (Tucson, AZ, 1982), pp. 1–2.

67 Geoffrey Blainey, *The Rise of Broken Hill* (Melbourne, 1968), p. 141.

68 Timothy Green, *The World of Diamonds* (London, 1981), p. 28.

69 Edward Jessup, *Ernest Oppenheimer: A Study in Power* (London, 1979), p. 98.

70 Jessup, *Ernest Oppenheimer*, pp. 110–13.

71 Ian M. Drummond, *The Gold Standard and the International Monetary System, 1900–1939* (Basingstoke, 1987), p. 44.

72 Schmitz, *World Non-Ferrous Metal Production and Prices*, pp. 90–91.

SIX · MASS PRODUCTION

1 Ronald Prain, *Reflections on an Era: Fifty Years of Mining in Changing Africa* (Worcester Park, Surrey, 1981), pp. 28–31.

2 N.J.G. Pounds & William N. Parker, *Coal and Steel in Western Europe* (Bloomington, IN, 1957), p. 242.

3 Alan Birch, *The Economic History of the British Iron and Steel Industry, 1784–1879* (London, 1967), pp. 378–84.
4 Pounds & Parker, *Coal and Steel in Western Europe*, p. 244.
5 'Copper in Germany', *Engineering and Mining Journal*, C/25 (25 December 1915), pp. 1056–8.
6 'Enemy Supplies of Metals and Ores', *Engineering and Mining Journal*, CIII/22 (2 June 1917), pp. 976–7.
7 Margaret L. Coit, *Mr Baruch* (Boston, 1957), pp. 164–9.
8 J. E. Spurr, *Political and Commercial Geology and the World's Mineral Resources* (New York, 1920), pp. 128–9.
9 Ray L. Wilbur and Arthur M. Hyde, *The Hoover Policies* (New York, 1937), p. 306.
10 Dan Rather, ed., *Our Times* (New York, 1996), p. 303.
11 William Y. Elliott, ed., *International Control in the Non-Ferrous Metals* (New York, 1937), pp. 710–22.
12 A. J. Wilson, *The Life and Times of Sir Alfred Chester Beatty* (London, 1985), pp. 203–7.
13 B.W.E. Alford and Charles E. Harvey, 'Copperbelt Merger: The Formation of the Rhokana Corporation, 1930–32', *Business History Review* (Autumn 1980), p. 187.
14 Charles C. Carr, *Alcoa: An American Enterprise* (New York, 1952), p. 114.
15 Winifred Lewis, *The Light Metals Industry* (London, 1949), pp. 134–5.
16 Barbara Tuchman, *The Guns of August* (New York, 1962).
17 Eric Schatzberg, *Wings of Wood, Wings of Metal* (Princeton, NJ, 1999), p. 38.
18 Philip Jarret, ed., *Monoplane to Biplane: Aircraft Development, 1919–1938* (London, 1997), p. 239.
19 Florence Hachez-Leroy, *L'Aluminium Français: L'Invention d'un Marche, 1911–1983* (Paris, 1984), p. 26.
20 Steven K. Holloway, *The Aluminium Multinationals and the Bauxite Cartel* (New York, 1988), p. 22.
21 George David Smith, *From Monopoly to Competition: The Transformations of Alcoa 1888–1986* (Cambridge, 1987), p. 107.
22 Donald H. Wallace, *Market Control in the Aluminium Industry* (Cambridge, 1937), p. 96.
23 Martin Fritz, *German Steel and Swedish Iron Ore, 1939–1945* (Göteberg, 1974), p. 17.
24 André Montagne, *Les Mines de Fer de Lorraine* (Nancy, 1991), p. 89.
25 'The Axis and Strategic Minerals', *Engineering and Mining Journal*, CXLIII/5 (1942), p. 42.
26 Lennard Bickel, *The Deadly Element: The Story of Uranium* (London, 1980), p. 55.
27 Earle Gray, *The Great Uranium Cartel* (Toronto, 1982), p. 28.
28 Thomas Powers, *Heisenberg's War: The Secret History of the German Bomb* (New York, 1993), p. 132.
29 M. E. Falkus, *The Industrialisation of Russia, 1700–1914* (London, 1972), p. 70.

30 K. H. Kennedy, *The Mining Czar* (London, 1986), pp. 64–70.
31 M. G. Corson, *Developments in the Non-Ferrous Metallurgical Industry and Science in Old Russia and the USSR* (New York, 1975), p. 56.
32 John D. Littlepage & Demaree Bess, *In Search of Soviet Gold* (London, 1947), p. 35.
33 Isaac Deutscher, *Stalin: A Political Biography* (London, 1961), p. 328.
34 Corson, *Developments ... in Old Russia and the USSR*, p. 62.
35 John Scott, *Behind the Urals: An American Worker in Russia's City of Steel* (Cambridge, MA, 1942), p. 64.
36 Stephen Kotkin, *Magnetic Mountain: Stalinism as a Civilisation* (Berkeley, CA, 1995), p. 43.
37 Littlepage & Bess, *In Search of Soviet Gold*, p. 109.
38 David Dallin & Boris Nicolaevsky, *Forced Labour in Soviet Russia* (New Haven, CT, 1947), p. 132.
39 Theodore Shabad, *Basic Industrial Resources of the USSR* (New York, 1969), p. 58.
40 Alfred E. Eckes, *The United States and the Global Struggle for Minerals* (Austin, TX, 1979), p. 210.
41 Kurt Meyer, *Pelletizing of Iron Ores* (Berlin, 1980), p. 7.
42 Jeffery A. Hart, *Rival Capitalists: International Competitiveness in the United States, Western Europe and Japan* (Ithaca, NY, 1992), p. 47.
43 L. Nasbeth and G. F. Ray, *The Diffusion of New Industrial Processes: An International Study* (Cambridge, 1974), pp. 146–58.
44 Eckes, *The United States and the Global Struggle*, p. 176.
45 Terutomo Ozawa, *Multinationalism, Japanese Style* (Princeton, NJ, 1979), p. 177.
46 John A. Howard, *Aerial Photo-Ecology* (New York, 1970), p. 4.
47 Roger Moody, *Plunder!* (London, 1991), p. 9.
48 John D. Wraith, *The Politics of Brazilian Development* (Stanford, CA, 1970), pp. 82–8.
49 Janet Henshall & R. P. Momsen, *A Geography of Brazilian Development* (London, 1976), pp. 128–9.
50 Raymond F. Mikesell, ed., *Foreign Investment in the Petroleum and Mineral Industries* (Baltimore, 1971), p. 332.
51 Gerald Manners, *The Changing World Market for Iron Ore, 1950–1980* (Baltimore, 1971), p. 253.
52 Tomohei Chida & Peter N. Davies, *The Japanese Shipping and Shipbuilding Industries* (London, 1990), p. 118.
53 'The 500 Largest Industrial Corporations', *Fortune Magazine* (May 1970), p. 184; 'The 200 Largest Industrial Companies Outside the U.S.', *Fortune Magazine* (August 1970), p. 143.
54 Kotkin, *Magnetic Mountain*, p. 29.
55 Paul A. Tiffany, *The Decline of American Steel* (New York, 1988), p. 65.
56 John Sheahan, *Promotion and Control of Industry in Post-War France* (Cambridge, 1963), p. 66.
57 William Diebold, Jr, *The Schuman Plan: A Study of Economic Co-operation 1950–59* (New York, 1959), p. 8.

58 Harry Schwartz, *The Soviet Economy since Stalin* (London, 1965), p. 126.
59 Ryutaro Komiya, ed., *Industrial Policy in Japan* (Tokyo, 1980), p. 281.
60 Seiichiro Yonekura, *The Japanese Iron and Steel Industry, 1850–1990* (New York, 1994), p. 212.
61 Ozawa, *Multinationalism, Japanese Style*, p. 164.
62 Marion Radetzki, 'The Role of State-owned Enterprises in the International Metal Mining Industry', *Resources Policy* (March 1989), p. 53.
63 Smith Hempstone, *Rebels, Mercenaries and Dividends: A History of the Belgian Congo* (New York, 1962), pp. 103–5.
64 James Dunkerley, *Rebellion in the Veins: Political Struggle in Bolivia* (London, 1984), p. 39.
65 Richard L. Sklar, *Corporate Power in an African State* (Berkeley, CA, 1975), p. 34.
66 Theodore H. Moran, *Multinational Corporations and the Politics of Dependence: Copper in Chile* (Princeton, NJ, 1974), p. 63.
67 Edy Kaufman, *Crisis in Allende's Chile: New Perspectives* (New York, 1988), p. 309.

List of Maps

Index